T0282381

Recent Advances in High-Temperature PEM Fuel Cells

Recent Advances in High-Temperature PEM Fuel Cells

Sivakumar Pasupathi, Juan Carlos Calderon Gomez,
Huaneng Su, Harikishan Reddy, Piotr Bujlo, Cordellia Sita

HySA Systems Competence Centre, South African Institute for Advanced
Materials Chemistry (SAIAMC), University of the Western Cape,
Cape Town, South Africa

Series Editor
Bruno G. Pollet

Power & Water, Swansea, Wales, UK

AMSTERDAM • BOSTON • HEIDELBERG • LONDON
NEW YORK • OXFORD • PARIS • SAN DIEGO
SAN FRANCISCO • SINGAPORE • SYDNEY • TOKYO

Academic Press is an imprint of Elsevier

Academic Press is an imprint of Elsevier
125 London Wall, London EC2Y 5AS, United Kingdom
525 B Street, Suite 1800, San Diego, CA 92101-4495, United States
50 Hampshire Street, 5th Floor, Cambridge, MA 02139, United States
The Boulevard, Langford Lane, Kidlington, Oxford OX5 1GB, United Kingdom

Copyright © 2016 Elsevier Ltd. All rights reserved.

No part of this publication may be reproduced or transmitted in any form or by any means,
electronic or mechanical, including photocopying, recording, or any information storage and retrieval
system, without permission in writing from the publisher. Details on how to seek permission, further
information about the Publisher's permissions policies and our arrangements with organizations
such as the Copyright Clearance Center and the Copyright Licensing Agency, can be found at our
website: www.elsevier.com/permissions

This book and the individual contributions contained in it are protected under copyright
by the Publisher (other than as may be noted herein).

Notices
Knowledge and best practice in this field are constantly changing. As new research and experience
broaden our understanding, changes in research methods or professional practices, may become
necessary.

Practitioners and researchers must always rely on their own experience and knowledge in evaluating
and using any information or methods described herein. In using such information or methods
they should be mindful of their own safety and the safety of others, including parties for whom they
have a professional responsibility.

To the fullest extent of the law, neither the Publisher nor the authors, contributors, or editors,
assume any liability for any injury and/or damage to persons or property as a matter of products
liability, negligence or otherwise, or from any use or operation of any methods, products, instructions,
or ideas contained in the material herein.

British Library Cataloguing-in-Publication Data
A catalogue record for this book is available from the British Library

Library of Congress Cataloging-in-Publication Data
A catalog record for this book is available from the Library of Congress

ISBN: 978-0-12-809989-6

For Information on all Academic Press publications
visit our website at https://www.elsevier.com

Working together
to grow libraries in
developing countries

www.elsevier.com • www.bookaid.org

Publisher: Joe Hayton
Acquisition Editor: Raquel Zanol
Editorial Project Manager: Mariana Kühl Leme
Production Project Manager: Sruthi Satheesh
Cover Designer: Victoria Pearson

Typeset by MPS Limited, Chennai, India

CONTENTS

Bruno G. Pollet is known in both academia and industry for his expertise and experience in the research fields of Hydrogen Energy, PEM Fuel Cell, Fuel Cell Electrocatalysis, Electrochemical Engineering and Sono(electro)chemistry. He is the Chief Technology Officer at Power and Water (KP2M Ltd), a young, dynamic and forward-thinking High Tech company in the UK. He is also Visiting Professor at the University of Ulster (UK) and the University of Yamanashi (Japan), after being a (full) Professor of Energy Materials and Systems at the University of the Western Cape (South Africa) and R&D Director of the National DST Hydrogen South Africa (HySA) Systems Competence Centre. He was also a co-founder and an Associate Director of the University of Birmingham Centre for Hydrogen and Fuel Cell Research (UK), and Operations and Delivery Director of the UK Engineering and Physical Sciences Research Council (EPSRC) Doctoral Training Centre in Hydrogen, Fuel Cells and their Applications.

Sivakumar Pasupathi is a Senior Lecturer and Programme Manager for HySA Systems Competence Centre, one of three national centres under the Hydrogen South Africa (HySA) programme, hosted by the South African Institute for Advanced Materials Chemistry (SAIAMC) at the University of the Western Cape (South Africa). He obtained his PhD from the University of Pisa (Italy) and has been working in the field of Hydrogen and Fuel Cells since late 1990s. He is an NRF rated Scientist and Principal Investigator in several bilateral projects with international partners. His current focus is on developing materials, components and systems, covering the whole value chain of Polymer Electrolyte Membrane (PEM) Fuel Cells, for both Combined Heat and Power and Fuel Cell Vehicle applications. His interests include PEM-based Fuel Cells and Electrolysers. He has published over 100 publications including papers in peer-reviewed international journals and conference proceedings, and he is an inventor of 11 patents and counting. He has spoken at several events and successfully organised international workshops and conferences.

CHAPTER *1*

Introduction

1.1 FUEL CELLS

A fuel cell is a device which converts chemical energy directly into electrical energy, electrochemically. The device operates at various temperatures (up to 1000°C), converting a fuel (e.g., hydrogen) and an oxidant (air or pure oxygen) in the presence of a catalyst into electricity, heat, and water. There are currently five major types of fuel cells; (1) the proton exchange membrane fuel cell (PEMFC) including (a) low-temperature PEMFCs (LT-PEMFCs), (b) high-temperature PEMFCs (HT-PEMFCs), and (c) direct methanol fuel cells (DMFCs), (2) the alkaline fuel cell (AFC), (3) the phosphoric acid fuel cell (PAFC), (4) the molten carbonate fuel cell (MCFC), (5) the solid oxide fuel cell (SOFC). A representative diagram of various types of Fuel Cells and their operating temperatures is given in Fig. 1.1 (Pollet et al., 2012).

1.2 FUEL CELLS IN THE FUTURE ENERGY MIX

The increase in global energy demand, along with increasing awareness of using greener and cleaner power sources, has placed an emphasis on renewable energy technologies to cater for our future energy needs. Owing to the intermittent nature of renewable sources, energy storage has become a key parameter in realizing the potential of this technology. Hydrogen and batteries are considered the most suitable renewable energy storage media in the long run. Renewable hydrogen coupled with fuel cells for generating electricity and heat is widely envisioned in an "ideal" green economy scenario. However, the current high-cost and durability issues of hydrogen and fuel cell technologies are hampering their mass deployment and commercialization.

Fuel cells have started to penetrate specific commercial markets, with some companies starting to see some profit (although little!) in this technology. Of the various types of fuel cells, polymer electrolyte membrane fuel cells (PEMFCs) stand out because of their flexibility in various applications. A PEMFC consists of a polymer electrolyte membrane

Recent Advances in High-Temperature PEM Fuel Cells. DOI: http://dx.doi.org/10.1016/B978-0-12-809989-6.00001-3
© 2016 Elsevier Ltd. All rights reserved.

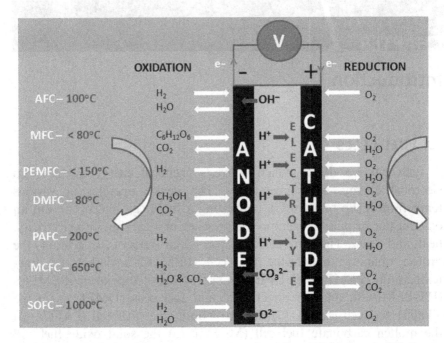

Figure 1.1 Types of fuel cells and their operating temperatures. Source: From Pollet, B.G., et al., 2012. Current status of hybrid, battery and fuel cell electric vehicles: From electrochemistry to market prospects. Electrochim. Acta 84, 235–249.

(PEM) sandwiched between two gas diffusion electrodes (GDEs) or two catalyst coated membranes. An efficient triple phase boundary is crucial for the PEMFC's performance and durability. Fig. 1.2 shows the different regions of a single-cell PEMFC. The LT-PEMFC is in the forefront of the drive to commercialization, particularly for stationary and transport applications, although LT-PEMFC has been under development since the 1960s.

1.3 HIGH-TEMPERATURE POLYMER ELECTROLYTE MEMBRANE FUEL CELLS

Although LT-PEMFCs are at the forefront of the commercialization of fuel cells, thanks to their high power density, rapid start-up and high efficiency, they suffer from many disadvantages. These include a low tolerance of fuel impurities (pure hydrogen (99.999%) or complex reformer systems with CO clean-up are required), low-quality heat generated, owing to the low operating temperatures and requirements for complex water management systems to prevent drying/flooding of

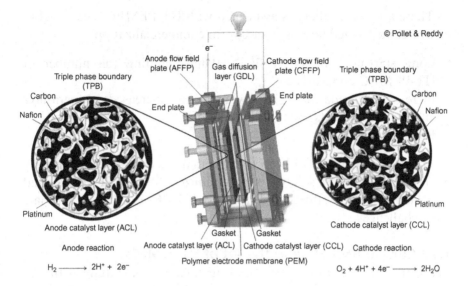

Figure 1.2 Exploded view of a single-cell PEMFC. Source: From http://www.sciencedirect.com/science/article/pii/B9781782423638000013.

the membrane electrode assembly (MEA). HT-PEMFCs, operating in the range 100–200°C, can overcome these issues, resulting in increased efficiency and simplification of the system.

1.3.1 Advantages and Disadvantages of HT-PEMFCs

The HT-PEMFC is a relatively new research area, which has attracted considerable interest recently (Rosli et al., in press). The higher operating temperatures offer several advantages:

- No need for water management systems
- High tolerance for fuel impurities (up to 3% CO in the fuel stream, which enables the use of a simple reformer system)
- High-quality heat that can be utilized for cogeneration purposes
- Simplified flow field plate (FFP) design due to improved transport of water (vapor) in the structures
- Minimized balance of plant (BoP) requirements enabling simpler system designs

The high tolerance of fuel impurities results in true fuel flexibility as conventional fuels (e.g., biogas, natural gas, methanol, or propane) can easily be converted into a hydrogen-rich gas, which can be fed directly to the fuel cell.

There are also challenges associated with HT-PEMFCs that need to be overcome to facilitate their successful commercialization:

- Long start-up time (typically >20 min) limiting the number of HT-PEMFC applications
- Increased degradation rates owing to the increased temperatures and the components used in this technology
- Relatively high cost

However, one needs to keep in mind that the components currently used for HT-PEMFCs have been optimized for LT-PEMFCs, and the ongoing R&D efforts dedicated to HT-PEMFCs may alleviate these challenges.

1.3.2 Current R&D State of the Art of HT-PEMFCs

R&D in HT-PEMFCs is gaining significant momentum with the number of publications increasing steadily. Most of the R&D is dedicated to the development of the PEM, although efforts are also being made to develop suitable catalysts and GDEs. Some of the R&D areas are as follows:

- Novel membranes that can operate between 40°C and 200°C without humidification requirements
- Development of suitable platinum group metal-based electrocatalysts that are active and durable
- Suitable supports for the catalysts, to overcome carbon support corrosion
- Suitable MEA architecture to reduce catalyst loading, minimize acid leaching and improve durability
- Improved FFP design and minimal FFP degradation
- Suitable stack designs with improved thermal management
- Improved system configurations and designs
- Minimized material degradation, including sealings and gaskets
- Diagnostic tools and protocols for performance and durability

1.3.3 Short-Term and Long-Term Market Drivers

The early-stage markets for HT-PEM are auxiliary power, battery chargers and combined heat and power (CHP) systems for the domestic and small-scale industrial sectors. Japan is at the forefront in domestic LT-PEM CHP installations (*Ene-Farm*) with over 140,000 fuel cell systems installed in 2015. Europe, particularly Germany, is following suit with domestic CHP installations, with its *Callux* and *Ene-Field* programs. HT-PEMFCs have a large role to play in this area owing to their fuel flexibility and efficiency for CHP applications. Regulations and government incentives are expected to play a significant part in promoting this technology. In the long-term HT-PEMFCs could also play a part as range extenders in electric vehicles and uninterruptible power supply systems. The main driver is the reduced system complexity and fuel flexibility, leading to better overall efficiency of HT-PEMFC systems.

1.3.4 Competing Technologies

LT-PEMFCs, HT-PEMFCs, and SOFCs are all being considered for stationary applications. Of the three technologies, SOFCs offer the highest electrical efficiency but their long start-up time and the need for continued operation during the night is a major issue. HT-PEMFCs are technically more suited for this application owing to better overall efficiency, sufficient dynamic operation and minimum BoP requirements. However, this requires an extensive R&D effort in order to fulfill the lifetime and cost requirements. Moreover, the "2020 targets for CHP fuel cell systems," >45% electrical efficiency, 90% energy efficiency, US$1000−1700/kW$_{avg}$ (depending on the size, kW$_{avg}$ is the average output of electric power delivered over the life of the system while it is running), start-up time of 20 min, and operating lifetime of 60,000 h, can potentially be achieved with HT-PEMFC systems.

REFERENCE

R.E. Rosli et al., A review of high temperature proton exchange membrane fuel cell (HT-PEMFC) system. Inter J Hyd Energy. In press.

CHAPTER 2

Catalysts for High-Temperature Polymer Electrolyte Membrane Fuel Cells

2.1 INTRODUCTION

Although there are several differences between high- and low-temperature polymer electrolyte membrane fuel cells, related to their operating conditions, the principles of catalyst design for these devices are essentially the same, and they are based upon two components: first, a suitable support, which favors the catalytic properties of the nanoparticles and promotes the electronic conductivity occurring on the catalyst sites. And secondly, nanoparticles with crystal distribution and size that make them active toward the redox reactions involved in these devices.

2.2 CATALYST SUPPORTS

Carbon black materials are the most employed materials, they are obtained by oil-furnace and acetylene processes, which are based on exothermic decomposition of oil fractions from petroleum and acetylene, respectively, generating materials with final ash content below 1 wt% (Antolini, 2009). This method consists of feeding the starting material into a furnace and burning it with a limited supply of air at about $1400°C$. At this temperature, radicals are formed and their recombination leads to the formation of carbon particles. The final characteristics of carbon black materials depend upon the process employed to make the tailoring of surface areas possible. For instance, carbon black materials from the acetylene process tend to have surface areas close to $50–90 \, m^2 \, g^{-1}$, whereas those from the oil-furnace process present areas higher than $200 \, m^2 \, g^{-1}$ (Antolini et al., 2002). Moreover, it is possible to recrystallize the carbon black material between $2500°C$ and $3000°C$, in order to enrich the carbon particles with well-ordered domains. The new material is known as graphitized carbon black (Antolini, 2009).

Recent Advances in High-Temperature PEM Fuel Cells. DOI: http://dx.doi.org/10.1016/B978-0-12-809989-6.00002-5
© 2016 Elsevier Ltd. All rights reserved.

The structure of carbon black is made up of parallel graphitic layers with a 0.35–0.38 nm interplanar distance, where carbon atoms have a sp^2 hybridization and form a triangular in-plane configuration. Moreover, atoms involved in the configuration possess p_z free orbitals able to form weaker delocalized π bonds with other carbon atoms (Peres et al., 2006). Another important characteristic of carbon blacks is the presence of surface functionalized groups, the most important being carboxylic acids, anhydrides, lactones and phenols. These oxygen groups increase the surface acidity of carbon, thereby increasing its adsorption ability toward basic molecules, and their incorporation is possible via oxidative treatments with nitric or sulfuric acids and potassium permanganate (Szymanski et al., 2002). Once these groups are present, they can be modified into nitrogen groups via reactions with nitrogen reagents (NH_3 and amines), which promotes basicity and improves interaction with acid molecules (Mangun et al., 2001). Nowadays, Ketjen, Vulcan, and acetylene black have been used as catalyst support materials in high-temperature polymer electrolyte membrane fuel cells (HT-PEMFCs) because of their high surface areas, the ease they afford of Pt nanoparticle loading, their relatively high electron conductivity and low cost (Kaewsai et al., 2015). Nevertheless, carbon black materials possess a large amount of 1-nm diameter micropores with a poor connection between them, avoiding the electroactive species diffusion (Antolini, 2009). Moreover, these materials are susceptible to corrosive (oxidative) conditions, particularly when the cathode potential is high and the pH is low (see Pourbaix diagram). Carbon is first oxidized, forming oxidized carbon intermediates (Eq. 2.1). The intermediates may be electroactive species susceptible to hydrolyzation (Eq. 2.2). The final step of the corrosion process is the formation of carbon dioxide (Eq. 2.3). The overall reaction is expressed as follows (Zhang et al., 2009):

$$C_s \rightarrow C_s^+ + e^- \tag{2.1}$$

$$C_s^+ + H_2O \rightarrow C_sO + 2H^+ + e^- \tag{2.2}$$

$$2C_sO + H_2O \rightarrow C_sO + CO_{2(g)} + 2H^+ + e^- \tag{2.3}$$

For these reasons, in the last few years, the use of novel synthetic carbon supports has attracted the attention of researchers looking for better properties such as high electronic conductivity and high mesopore content—mesopores play a key role in the activity of

Table 2.1 Textural and Morphological Properties of Some Typical Carbon Supports for PEMFCs

Carbon Support	Surface Area $(m^2\,g^{-1})$	Pore Volume $(cm^3\,g^{-1})$	Micropores Volume $(cm^3\,g^{-1})$	Mesopores Volume $(m^3\,g^{-1})$	Micropores Area $(m^2\,g^{-1})$	Mesopores Area $(m^2\,g^{-1})$	References
Vulcan XC-72 R carbon black	218	0.41	0.036	0.37	65	153	Sebastian (2010)
Carbon nanofibers	185	0.69	0.01	0.68	23.9	161	Sebastian (2010)
Carbon nanotubes	415	0.8	0.11	0.67	210	205	Cheng and Jiang (2013)
Carbon microspheres	1280	0.75	0.52	0.24	1060	220	Yang et al. (2005)
Carbon xerogels	671	0.893	0.177	0.716	0.1	274	Calderón et al. (2012)
Ordered mesoporous carbons	577	0.37	0.07	0.3	133	444	Calvillo et al. (2007a,b)

catalysts. Some of these novel carbon materials are carbon nanofibers (CNF, Sebastián et al., 2010), carbon nanotubes (CNT, Cheng and Jiang, 2013), carbon microspheres, hard carbon spherules (Yang et al., 2005), carbon aerogels and xerogels (Calderón et al., 2012) and ordered mesoporous carbons (Calvillo et al., 2007a). Textural and morphological properties of these materials are reported in Table 2.1, where it is possible to see that some of the synthetic carbons possess higher surface areas and mesoporous volumes than those observed for the Vulcan XC-72R carbon black. These are the two most important properties to take into account in carbon supports for polymer electrolyte membrane fuel cells' (PEMFCs') catalysts, since they are directly related to the dispersion of catalytic nanoparticles and the diffusion of the electroactive species. For further information on support materials for fuel cell catalysts, please refer to the excellent review by Sharma and Pollet (2012).

As mentioned earlier, in the case of HT-PEMFCs, corrosion of carbon support is enhanced compared to low-temperature polymer electrolyte membrane fuel cells (LT-PEMFCs), and becomes an important factor, since it induces the disturbance of pore structure, modifying both the electroactive species' transport to the catalyst-active sites and the removal of residues. These changes can significantly contribute deterioration in fuel cell performance (Schulenburg et al., 2011). Thus, it is imperative to use materials exhibiting resistance to high temperatures, when applied potentials are higher than +1 V, such as carbon nanotubes (CNTs, cylindrical graphene tubes of nanosized diameters). It seems to be one of the carbon materials most frequently employed as support for catalysts in HT-PEMFCs, which corresponds to the findings of some recent studies reported in the literature (Kaewsai et al., 2015; Okamoto et al., 2009; Berber et al., 2014), owing to its corrosion resistance, outstanding electrochemical durability, notable electron and thermal conductivity, high mechanical strength, good chemical and thermal stability and large surface areas. Moreover, its cross-linked fibers in the catalyst layer promote the diffusion of fuel gas (Tian et al., 2006).

However, CNTs lack defect sites, an important morphological property for metal binding and durability. As in the case of other carbon supports, this issue is solved with strong acid treatments that provide defect sites and also form surface-active functional groups such as $-COOH$, $-CHO$, $=O$, and $-OH$, all of them able to promote the metal precursors anchoring (Wepasnick et al., 2011). Table 2.2 shows the types and content of some surface functional

Table 2.2 Chemical Treatments Applied to Some Carbon Supports, and Surface Functional Groups Generated Afterwards

Carbon Support	Funtionalization Procedure	Functional Group Content (mmol g^{-1})					References
		Carboxylic Acids	Anhydride Lactone	Anhydride	Phenol	Lactone	
Carbon nanofibers	Diluted nitric acid	1	3.75	0	14	16.84	Calvillo et al. (2007a,b)
	Concentrated nitric acid	3.17	6.43	0	9.98	31.78	
	Nitric–sulfuric acid mixture	6.42	13.35	0.16	19.93	33.53	
Ordered mesoporous carbons	Diluted nitric acid	47.4	12.8	14.9	146.1	133.1	Lazaro et al. (2007)
	Concentrated nitric acid	134.2	53.7	51.9	247.1	80.1	
	Nitric–sulfuric acid mixture	92.2	112.1	36	247.4	125.1	
Carbon xerogels	Diluted nitric acid	0.39	0.24	0.09	0.17	0.64	Alegre et al. (2012)
	Concentrated nitric acid	0.77	0.35	0.3	1.95	1.05	
	O_2–N_2 mixture	0.06	0.56	0	0.91	0.08	

groups obtained after various chemical treatments have been applied to some carbon support materials (Calvillo et al., 2007a,b; Lazaro et al., 2007; Alegre et al., 2012).

An interesting and widely reported alternative to chemical modification is the wrapping of carbon supports with polymers, such as polyallylamine hydrochloride (Zhang et al., 2010), polyaniline (Yoo et al., 2011), poly(diallyldimethylammonium chloride) (Zhang et al., 2011), and polypyrrole (Oh et al., 2011). These polymers promote the impregnation of metal precursors during synthesis via the interaction with attractive forces such as *van der Waals* forces, electrostatic forces, or hydrogen bonding. Especially, $\pi-\pi$ electrons in CNTs are highly compatible with some *Lewis base* electron-donor functional groups such as $-N:$, $-O:$, and $-S:$ from the polybenzimidazole (PBI) and pyridine-modified polybenzimidazole (PyPBI) (Okamoto et al., 2009). CNTs wrapped with PyPBI acting as supports for cathode Pt catalysts exhibit higher durability at high-operating fuel cell temperatures than those reported for carbon blacks—PyPBI composites (and also surpass the performances reported for MEAs with a commercial catalyst Pt/C on the cathode side (Berber et al., 2014)).

2.3 CATALYSTS FOR HT-PEMFCs

Pure platinum (Pt) supported on a suitable support material, is the metal most frequently used as a catalytic phase for HT-PEMFC electrodes, as well as for LT-PEMFC electrodes. Once hydrogen and oxygen are transported through the electrolyte to the active area of the catalyst nanoparticles, they react in the three phase boundary between the electrolyte, the electroactive species and the catalyst nanoparticles. The transport resistance of reactants, the kinetics of electron transfer, and proton transport in the catalyst layer are factors that are important in influencing reaction rates for these processes (Authayanun et al., 2015). In order to decrease the level of use of Pt as catalyst, bearing in mind its high cost and low availability, and also the necessity to overcome some typical problems presented in these devices (i.e., CO poisoning and the sluggish kinetics of oxygen reduction), the implementation and use of some Pt alloys with transition metals have been proposed. In the case of the anode, some of these metals are Ru, Sn, Ni, and Rh. For the cathode, the typical metals are Co, Cr, Fe, Cu, and Pd, (Stamenkovic et al., 2007). All of these alloys are prepared by several synthesis procedures such as sodium borohydride reduction, formic acid, sodium formate, methanol, or polyols (Calderón et al., 2015).

Nevertheless, the use of Pt alloys is less common when PEMFCs are operated at high temperatures, because in these devices the increase of temperature is useful to overcome the above-mentioned issues present in LT-PEMFCs. A clear example of this behavior is CO poisoning, a typical issue that occurs in the anode when the hydrogen originates from hydrocarbon reformate processes (Zhang et al., 2006). The electrochemical oxidation of hydrogen in PEM fuel cells is addressed by its dissociative adsorption, which requires free platinum catalyst sites to yield two protons (H^+) and two electrons (hydrogen oxidation reaction). There is a large exchange current density (i_o) with low overpotentials ($\eta < +100mV$) associated with this process and thus, a fast reaction rate is involved. When CO is present in the H_2 stream, it is adsorbed on the Pt active site, inhibiting hydrogen oxidation by means of its link via linear bonding and bridge bonding, occupying one and two adsorption platinum surface sites, respectively (Li et al., 2003). If the operating temperature is increased, the CO amount adsorbed on the Pt surface is reduced, because of the high negative value for CO adsorption standard entropy and as a consequence more free sites on the catalyst surface for the H_2 adsorption are formed (Dhar et al., 1986).

In the case of the cathode, the oxygen reduction reaction (ORR) is also affected by the increase of temperature in HT-PEMFCs. At low temperatures, it is well known that the kinetics for this reaction is sluggish and thus, a decrease in LT-PEMFC performance is observed. As already mentioned, for several years the generally accepted solution to this problem has been the design of Pt alloys with the addition of some transition metals such as Pd, Co, Cr, Ni, Fe, and Cu, and several studies have reported the enhancement of activity in pursuit of ORR with the use of these alloys as cathode catalysts (Jung et al., 2012). Recently, other alternatives, such as transition metal porphyrin complexes and metal oxides have also been suggested (Olson et al., 2008). Certainly, the implementation of these new catalysts has involved the use of a low Pt loading at the cathode, and therefore, a drop in production costs. Nonetheless, at high temperatures, the kinetics of oxygen reduction is expected to be increased; however, owing to the presence of phosphoric acid, which blocks some of the reactant sites, an increased platinum loading is required compared to those in LT-PEMFCs (Su et al., 2013).

The drop in the activity of cathodes in HT-PEMFCs is basically caused by two factors: low oxygen permeability at high temperatures and strong phosphate adsorption, the latter being the one which produces the most damaging effects. The presence of phosphate anions comes from the use of H_3PO_4-doped PBI membranes as a replacement for Nafion-based membranes, considering their better proton conductivity at high temperatures, in comparison with the nondoped membranes. Some rotating disk electrode studies related to the adsorption of phosphate ions on Pt single crystal electrodes demonstrated that these anions can strongly adsorb onto the Pt surface and that this in turn decreases the kinetics of the ORR (He et al., 2010). Moreover, corrosion in the presence of H_3PO_4 is promoted, inducing the sintering and electrical isolation of Pt nanoparticles, as well as the enhancement of carbon support hydrophilicity, which modifies the mass-transport in the gas diffusion electrode (Bauer et al., 2012).

A way to overcome these issues is the replacement of conventional carbon supports with perovskite-like compounds such as lanthanum manganite (Hayashi et al., 2004), although a considerable disadvantage of these materials is their low electronic conductivity. Other authors suggested the use of Pt-metal oxide catalysts, which have been shown to exhibit higher levels of activity than those observed for Pt catalysts. Some examples of these materials are $PtWO_3$, $PtMnO_2$, $PtCrO_2$, and PtV_2O_5, all of them supported on a carbon support (Tseung and Chen, 1997). Combining Pt with zirconium oxides could be another interesting alternative for cathode catalysts in HT-PEMFCs. Liu et al. (2007) prepared a Pt_4ZrO_2 catalyst supported on carbon black, using a mixture of a commercial Pt/C catalyst and the previously prepared ZrO_2/C material. They found a significant stability after 3000 cycles for the MEA built with Pt_4ZrO_2/C, in comparison with the results observed for a commercial Pt/C catalyst MEA. Power densities for the Pt_4ZrO_2/C MEA were also higher than those observed for the Pt/C MEA and these results were attributed to the low sintering temperature displayed by the Pt_4ZrO_2/C catalyst, as concluded by means of the TEM images analyses made for the employed materials, before and after the tests.

Tin (Sn) has also been used in the design of cathode catalysts for HT-PEMFCs. Parrondo et al. (2010) synthesized a Pt catalyst supported on a previously prepared SnO_x/C composite. From the polarization curves obtained using PBI-based MEAs, they observed the lowest potential drops with the smallest SnO_x concentration in the catalyst support (7 wt%). High concentrations of tin oxide reduced the performance of the fuel cell, and this observation was explained by the mass transport limitations within the electrode, although in all cases, the presence of Sn oxides produced lower potential drops than those observed for a commercial Pt/C catalyst. TiO_2-based materials as support materials for Pt catalysts are another promising alternative, owing to their high thermal and chemical stability, exceptional corrosion resistance, good proton conductivity and good interactions between the catalyst particles and the TiO_2 support (Antolini and Gonzalez, 2009). For instance, Bauer et al. (2012) reported the synthesis of catalysts supported on Nb-doped TiO_2 mesoporous microspheres and nanofibers, in order to evaluate their stability in acidic media. Electrochemical characterizations of the prepared materials showed no significant changes in the typical electrochemical signals after 1000 scan cycles at $100 \, mV \, s^{-1}$ for the Pt/Nb-TiO_2 materials, a behavior explained from the improved corrosion resistance of Nb-TiO_2 ceramic support, in comparison with the notoriously low corrosion resistance of carbon materials. In fact, the catalyst supported on carbon black, which was employed for comparison purposes, displayed a decrease in the current density values for the hydrogen adsorption–desorption process and oxide formation signals, after 100 operation cycles. In spite of these positive results, the authors suggested an improvement in conductivity for these supports, in order to increase the mass transport-limited current obtained from the ORR.

In the case of Pt alloys with transition metals, some works suggested the use of cobalt in cathodic catalysts for HT-PEMFCs. Schenk et al. (2014) synthesized Pt-Co alloys supported directly on the porous layer of a gas diffusion electrode, in order to avoid the conventional steps of filtering, centrifugation, washing, drying, and annealing involved in the typical synthesis procedures. Moreover, the effect of an acid leaching on the performance of these catalysts was evaluated; this treatment was applied in order to remove the Co residues present on the catalyst surface, which were not successfully reduced during the electrode preparation. The acid-leached and heat-treated electrode displayed the best definition in the hydrogen adsorption–desorption signals, as well as in those corresponding for the Pt oxides formation, even after cycling for 600 h. The polarization and power density curves showed an in single-cell performance when these materials were employed as cathodes, especially after 14 h of operation, a behavior attributed to a better distribution of the electrolyte within the catalyst layer and the cleaning of the catalyst's surface. Table 2.3 presents the values for some electrochemical parameters obtained after the characterization of the cited catalysts, compared with some commercial Pt/C catalysts.

From the studies cited above, it is possible to conclude that corrosion is the principal problem in HT-PEMFCs, therefore the design of the catalysts for these devices demands the use of resistant materials. In this sense, CNTs have been recognized as an "ideal" support for this type of application, because of their well-known good electrical and morphological properties. However, they require some polymer-wrapping treatments in order to expedite the anchoring of metal precursors during synthesis procedures, PBI and PyPBI being the most commonly employed, because of their strong interactions with CNTs. The main result of this wrapping process has been the increase of catalyst durability after several operating hours.

2.4 SUMMARY

Enhancement of temperature for HT-PEMFCs has allowed some common technical problems present in the low-temperature proton exchange membrane fuel cell to be overcome. In the case of the anode, CO adsorption on the catalytic phase is low and the use of Pt-transition metals alloys is not imperative, especially considering the high sensitivity of transition metals to corrosion at high temperatures. As a consequence, only Pt is used in these anodes. In the case of the cathode, increase of kinetics for the ORR is not evident owing to the adsorption of phosphate anions, which are mainly produced from the use of H_3PO_4-doped polybenzimidazole membranes. Therefore, the implementation and use of Pt-composites with metals more resistant to corrosion, such as Zr, Sn, and Ti, has been suggested. Some reports

Table 2.3 Electrochemical Performance of Some Catalysts Employed in HT-PEMFCs

Catalyst	Electrochemical Performance					References
	Current Density (Before Cycling Treatment) (mA cm^{-2})	Power Density (Before Cycling Treatment) (mW cm^{-2})	Current Density (After Cycling Treatment) (mA cm^{-2})	Power Density (After Cycling Treatment) (mW cm^{-2})		
Pt$_4$ZrO$_2$/C	4000	1160	3750	900		Liu et al. (2007)
Pt/C TKK 47%	3500	1000	2500	660		
	SnO Concentration (wt%)	OCV (V)	ECA (m^2 g^{-1} Pt)	Crossover Current (mA cm^{-2})		
Pt/SnO$_x$/C	0	0.95	3.7	1.8		Parrondo et al. (2010)
	7	0.85	4.9	2.7		
	12	0.85	5	2.8		
	18	0.93	2	0.3		
	24	0.77	4.3	2.6		
	Support Type	ECSA After 100 Cycles (m^2 g^{-1} Pt)	ECSA After 500 Cycles (m^2 g^{-1} Pt)	ECSA After 1000 Cycles (m^2 g^{-1} Pt)	Mass Activity at 0.9 V vs RHE (A g^{-1} Pt)	
Pt/Nb-doped TiO$_2$	TiO$_2$ mesoporous microspheres	25	25	25	3.9	Bauer et al. (2012)
	TiO$_2$ nanofibers	22	22	22	3.2	
	Commercial carbon	48	36	25	9.2	
	SA (mA cm^{-2})	MA (A g^{-1} Pt)	Initial ECSA (cm^2 mg^{-1} Pt)	ECSA Loss After Long-Term Operation (%)		
Pt-Co/C untreated	0.09	0.018	198	38		Schenk et al. (2014)
Pt-Co/C leached	0.318	0.018	257	23		
Pt-Co/C leached + HT	0.322	0.1	310	20		
Pt/C	0.188	0.049	260	49		

have shown that the use of zirconium precludes the sintering of Pt nanoparticles, enhancing the stability of MEAs after long operating times. The presence of tin oxides in the cathodes minimizes cell voltage drops in the polarization curves, although a high concentration of these materials hinders the mass transport and reduces the performance of the electrodes. Finally, the use of TiO_2 increases the MEAs' durability after several operating periods, particularly if the composite Pt-TiO_2 is doped with Niobium. In spite of the well-known high corrosion resistance of Ti compounds, it is necessary to improve the conductivity of this material. In general, research in catalyst design for high-temperature proton exchange membrane fuel cells must be improved in order to obtain low-cost materials able to overcome the typical technical problems present in these devices, producing the maximum power and current densities.

REFERENCES

Alegre, C., Galvez, M.E., Baquedano, E., Pastor, E., Moliner, R., Lazaro, M.J., 2012. Influence of support's oxygen functionalization on the activity of Pt/carbon xerogels catalysts for methanol electro-oxidation. Int. J. Hydrogen Energy 37 (8), 7180–7192.

Antolini, E., 2009. Carbon supports for low temperature fuel cells catalysts. Appl. Catal. B Environ. 88 (1–2), 1–24.

Antolini, E., Gonzalez, E.R., 2009. Ceramic materials as supports for low temperature fuel cell catalysts. Solid State Ionics 180 (9–10), 746–763.

Antolini, E., Passos, R.R., Ticianelli, E.A., 2002. Effects of the carbon powder characteristics in the cathode gas diffusion layer on the performance of polymer electrolyte fuel cells. J. Power Sources 109 (2), 477–482.

Authayanun, S., Im-orb, K., Arpornwichanop, A., 2015. A review of the development of high temperature proton exchange membrane fuel cells. Chinese J. Catal. 36 (4), 473–483.

Bauer, A., Chevallier, L., Hui, R., Cavaliere, S., Zhang, J., Jones, D., et al., 2012. Synthesis and characterization of Nb-TiO_2 mesoporous microsphere and nanofiber supported Pt catalysts for high temperature PEM fuel cells. Electrochim. Acta 77, 1–7.

Berber, M.R., Hafez, I.H., Fujigaya, T., Nakashima, N., 2014. Durability analysis of polymer-coated pristine carbon nanotube-based fuel cell electrocatalysts under non-humidified conditions. J. Mater. Chem. A 2, 19053–19059.

Calderón, J.C., Mahata, N., Pereira, M.F.R., Figueiredo, J.L., Fernandes, V.R., Rangel, C.M., et al., 2012. Pt-Ru catalysts supported on carbon xerogels for PEM fuel cells. Int. J. Hydrogen Energy 37 (8), 7200–7211.

Calderón, J.C., García, G., Calvillo, L., Rodríguez, J.L., Lázaro, M.J., Pastor, E., 2015. Electrochemical oxidation of CO and methanol on Pt–Ru catalysts supported on carbon nanofibers: the influence of synthesis method. Appl. Catal. B Environ. 165, 676–686.

Calvillo, L., Lázaro, M.J., Bordejé, E.G., Moliner, R., Cabot, P.L., Esparbé, I., et al., 2007a. Platinum supported on functionalized ordered mesoporous carbon as electrocatalysts for direct methanol fuel cells. J. Power Sources 169 (1), 59–64.

Calvillo, L., Lázaro, M.J., Suelves, I., Echegoyen, Y., Bordejé, E.G., Moliner, R., 2007b. Study of the surface chemistry of modified carbon nanofibers by oxidation treatments in liquid phase. J. Nanosci. Nanotechnol. 9 (7), 4164–4169.

Cheng, Y., Jiang, S.P., 2013. Highly effective and CO-tolerant PtRu electrocatalysts supported on poly(ethyleneimine) functionalized carbon nanotubes for direct methanol fuel cells. Electrochim. Acta 99, 124–132.

Dhar, H.P., Christner, L.G., Kush, A.K., Maru, H.C., 1986. Performance study of a fuel cell Pt-on-C anode in presence of CO and CO_2 and calculation of adsorption parameters for CO poisoning. J. Electrochem. Soc. 133 (8), 1574–1582.

Hayashi, M., Uemura, H., Shimanoe, K., Miura, N., Yamazoe, N., 2004. Reverse micelle assisted dispersion of lanthanum manganite on carbon support for oxygen reduction cathode. J. Electrochem. Soc. 151 (1), A158–A163.

He, Q., Yang, X., Chen, W., Mukerjee, S., Koel, B., Chen, S., 2010. Influence of phosphate anion adsorption on the kinetics of oxygen electroreduction on low index Pt(hkl) single crystals. Phys. Chem. Chem. Phys. 12, 12544–12555.

Jung, G.B., Tseng, C.C., Yeh, C.C., Lin, C.Y., 2012. Membrane electrode assemblies doped with H_3PO_4 for high temperature proton exchange membrane fuel cells. Int. J. Hydrogen Energy 37 (18), 13645–13651.

Kaewsai, D., Lin, H.L., Liu, Y.C., Yu, T.L., 2015. Platinum on pyridine-polybenzimidazole wrapped carbon nanotube supports for high temperature proton exchange membrane fuel cells. Int J Hydrogen Energy. Available from: http://dx.doi.org/10.1016/j.ijhydene.2015.09.066.

Lazaro, M.J., Calvillo, L., Bordeje, E.G., Moliner, R., Juan, R., Ruiz, C.R., 2007. Functionalization of ordered mesoporous carbons synthesized with SBA-15 silica as template. Microporous Mesoporous Mater. 103 (1–3), 158–165.

Li, Q.F., He, R.H., Gao, J.A., Jensen, J.O., Bjerrum, N.J., 2003. The CO poisoning effect in PEMFCs operational at temperatures up to 200°C. J. Electrochem. Soc. 150 (12), A1599–A1605.

Liu, G., Zhang, H., Zhai, Y., Zhang, Y., Xu, D., Shao, Z., 2007. Pt_4ZrO_2/C cathode catalyst for improved durability in high temperature PEMFC based on H_3PO_4 doped PBI. Electrochem. Commun. 9 (1), 135–141.

Mangun, C.L., Benak, K.R., Economy, J., Foster, K.L., 2001. Surface chemistry, pore sizes and adsorption properties of activated carbon fibers and precursors treated with ammonia. Carbon 39 (12), 1809–1820.

Oh, H.S., Kim, K., Kim, H., 2011. Polypyrrole-modified hydrophobic carbon nanotubes as promising electrocatalysts supports in polymer electrolyte membrane fuel cells. Int. J. Hydrogen Energy 36 (18), 11564–11571.

Okamoto, M., Fujigaya, T., Nakashima, N., 2009. Design of an assembly of poly(benzimidazole), carbon nanotubes, and Pt nanoparticles for a fuel-cell electrocatalyst with an ideal interfacial nanostructure. Small 5 (6), 735–740.

Olson, T.S., Chapman, K., Atanassov, P., 2008. Non-platinum cathode catalyst layer composition for single membrane electrode assembly proton exchange membrane fuel cell. J. Power Sources 183 (2), 557–563.

Parrondo, J., Mijangos, F., Rambabu, B., 2010. Platinum/tin oxide/carbon cathode catalyst for high temperature PEM fuel cell. J. Power Sources 195 (13), 3977–3983.

Peres, N.M.R., Guinea, F., Castro Neto, A.H., 2006. Electronic properties of two-dimensional carbon. Ann. Phys. 321 (7), 1559–1567.

Schenk, A., Grimmer, C., Perchthaler, M., Weinberger, S., Pichler, B., Heinzl, C., et al., 2014. Platinum-cobalt catalysts for the oxygen reduction reaction in high temperature proton exchange membrane fuel cells—long term behavior under ex-situ and in-situ conditions. J. Power Sources 266, 313–322.

Schulenburg, H., Schwanitz, B., Linse, N., Scherer, G.G., Wokaun, A., 2011. 3D imaging of catalyst support corrosion in polymer electrolyte fuel cells. J. Phys. Chem. C 115 (29), 14236–14243.

Sebastián, D., Calderón, J.C., González-Expósito, J.A., Pastor, E., Martínez-Huerta, M.V., Suelves, I., et al., 2010. Influence of carbon nanofiber properties as electrocatalyst support on the electrochemical performance for PEM fuel cells. Int. J. Hydrogen Energy 35 (18), 9934–9942.

Sharma, S., Pollet, B.G., 2012. Support materials for PEMFC and DMFC electrocatalysts—a review. J. Power Sources 208 (15), 96–119.

Stamenkovic, V.R., Fowler, B., Mun, B.S., Wang, G.F., Ross, P.N., Lucas, C.A., et al., 2007. Improved oxygen reduction activity on $Pt_3Ni(111)$ via increased surface site availability. Science 315 (5811), 493–497.

Su, H.N., Pasupathi, S., Bladergroen, B., Linkov, V., Pollet, B.G., 2013. Optimization of gas diffusion electrode for polybenzimidazole-based high temperature proton exchange membrane fuel cell: evaluation of polymer binders in catalyst layer. Int. J. Hydrogen Energy 38 (26), 11370–11378.

Szymanski, G.S., Karpinski, Z., Biniak, S., Swiatkowski, A., 2002. The effect of the gradual thermal decomposition of surface oxygen species on the chemical and catalytic properties of oxidized activated carbon. Carbon 40 (14), 2627–2639.

Tian, Z.Q., Jiang, S.P., Liang, Y.M., Shen, P.K., 2006. Synthesis and characterization of platinum catalysts on multiwalled carbon nanotubes by intermittent microwave irradiation for fuel cell applications. J. Phys. Chem. B 110 (11), 5343–5350.

Tseung, A.C.C., Chen, K.Y., 1997. Hydrogen spill-over effect on Pt/WO_3 anode catalysts. Catal. Today 38 (4), 439–441.

Wepasnick, K.A., Smith, B.A., Schrote, K.E., Wilson, H.K., Diegelmann, S.R., Fairbrother, D. H., 2011. Surface and structural characterization of multiwalled carbon nanotubes following different oxidative treatments. Carbon 49 (1), 24–36.

Yang, R., Qiu, X., Zhang, H., Li, J., Zhu, W., Wang, Z., et al., 2005. Monodispersed hard carbon spherules as a catalyst support for the electrooxidation of methanol. Carbon 43 (1), 11–16.

Yoo, S.J., Kim, S.K., Jeon, T.Y., Hwang, S.J., Lee, J.G., Lee, S.C., et al., 2011. Enhanced stability and activity of Pt-Y alloy catalysts for electrocatalytic oxygen reduction. Chem. Commun. 47, 11414–11416.

Zhang, J.L., Xie, Z., Zhang, J.J., Tang, Y.H., Song, C.J., Navessin, T., et al., 2006. High temperature fuel cells. J. Power Sources 160 (2), 872–891.

Zhang, S., Yuan, X., Wang, H., Mérida, W., Zhu, H., Shen, J., et al., 2009. A review of accelerated stress tests of MEA durability in PEM fuel cells. Int. J. Hydrogen Energy 34 (1), 388–404.

Zhang, S., Shao, Y.Y., Yin, G.P., Lin, Y.H., 2010. Carbon nanotubes decorated with Pt nanoparticles via electrostatic self-assembly: a highly active oxygen reduction electrocatalyst. J. Mater. Chem. 20, 2826–2830.

Zhang, S., Shao, Y.Y., Yin, G.P., Lin, Y.H., 2011. Self-assembly of Pt nanoparticles on highly graphitized carbon nanotubes as an excellent oxygen-reduction catalyst. Appl. Catal. B Environ. 102 (3–4), 372–377.

CHAPTER 3

Advances in HT-PEMFC MEAs

3.1 SELECTION OF THE BINDER FOR HT-MEA

At present, almost all membrane electrode assemblies (MEAs) for polybenzimidazole (PBI)-based high-temperature proton exchange membrane fuel cell (HT-PEMFC) are prepared by the catalyst-coated gas diffusion layer (GDL) method. In other words the MEA is the assembly of an acid-doped membrane sandwiched between two gas diffusion electrodes (GDEs). Therefore, the GDE characteristics, especially the structure of the catalyst layer (CL), a thin layer on the GDL, have significant influence on the overall fuel cell performance. Usually, the CL consists of two components: the catalyst particles and the polymer binder. The polymer binder plays an important role in determining the GDE performance. One other aspect is the binder content. If the content is too high, gas diffusion becomes problematic. But if the content is too low, ion conductivity suffers and more of the catalyst becomes isolated. On the other hand, binder properties also have an effect on the GDE performance, e.g., in terms of interfacial contact and durability, owing to their different affinities with the electrolyte membrane and the electrolyte polymer in the CL. A few researchers have performed investigations in this field and thus it is difficult to decide which binder to use for HT-MEAs.

Park et al. (2011) studied the properties of binders (polytetrafluoroethylene (PTFE), polyurethane, and PBI) in the CLs of GDEs used in HT-MEA and their influences on both the HT-PEMFC performance and lifetime. They found that the cell voltage at $0.3\,A\,cm^{-2}$ varied from $+0.583$ to $+0.619\,V$ by changing the binder material from PTFE to polyurethane. By using binders such as PBI and polyurethane in the GDEs, *Tafel* slope (*b*) value of $+87\,mV\,dec^{-1}$ was found, while the electrodes with PTFE binders resulted in a *Tafel* slope of $+100\,mV\,dec^{-1}$. The electrodes with polyurethane binders showed Pt utilization and η_{mass} values close to the electrodes with PTFE binders; however, the electrodes with PBI binders showed lower Pt

Recent Advances in High-Temperature PEM Fuel Cells. DOI: http://dx.doi.org/10.1016/B978-0-12-809989-6.00003-7
© 2016 Elsevier Ltd. All rights reserved.

utilization and higher η_{mass} values. Thus the electrode with PBI binders did not show any improvement in fuel cell performances. The change in *Tafel* slopes contributed to the high cell voltage for the electrodes that used polyurethane binders.

Also, Mazúr et al. (2011) studied the nature and optimized the content of the polymeric binder (PTFE—hydrophobic, PBI—hydrophilic) in the CL and concentration of Pt in catalytic powder (affecting the thickness of the CL). From the results obtained in their study, it was observed that both PBI and PTFE can be used as binders in the CL. However, when hydrophilic PBI was used as the binder, the risk of the CL flooding by phosphoric acid (PA) was significantly high. In the case of PTFE, the danger of such behavior was relatively low with a sufficiently thick CL (30 μm or more). For the thinner CL, sufficiently high-PTFE content has to be used to prevent the thin CL from PA flooding.

Recently, Su et al. (2013) fully evaluated the most currently used polymer binders (Nafion, PBI, PTFE, PVDF, and PBI-PVDF blends) with a view to optimizing the GDE performance of PBI-based HT-PEMFCs. They confirmed that conventional Nafion and PBI polymers are less attractive as binders and ionic conductors in HT-PEMFC CLs. They found that PTFE and polyvinylidene difluoride (PVDF) are preferred polymer binders for high-performance MEAs for HT-PEMFC, owing to the superior CL structures and electrochemical properties. Almost all MEAs for PBI- or ABPBI-based HT-PEMFCs were prepared with either PTFE or PVDF binders.

3.2 TECHNIQUES USED FOR HT-MEA FABRICATION

Most of the R&D in HT-MEAs is dedicated to the development of the membrane, Table 3.1, and the most common HT membrane structures are given in Table 3.2. Efforts are also being made to develop suitable catalysts, catalyst support materials, and GDEs.

The focus on MEA architecture is aiming at reducing catalyst loading and improving durability. Various coating techniques have been developed and used for both catalyst-coated substrate (CCS) and catalyst-coated membrane (CCM) MEA preparation methods. A few examples of the preparation methods are shown in Table 3.3 (Chandan et al., 2013).

Table 3.1 List of HT Membranes and Their Properties

Membrane Material	Conductivity/Temp/%RH	References
Nafion/(5wt%) SPPSQ*	0.157 S cm^{-1} at 120°C and 100% RH	Nam et al. (2008)
PBI/H$_3$PO$_4$/(40%) SiWA	0.177 S cm^{-1} at 150°C and 0% RH	Verma et al. (2010)
Recast Nafion**	0.002 S cm^{-1} at 130°C and 100% RH	Yuan et al. (2009)
Nafion*	0.035 S cm^{-1} at 120°C and 30% RH	Zarrin et al. (2011)
Nafion 117/(20 wt%) ZrSPP	0.1 S cm^{-1} at 100°C and 90% RH	Casciola et al. (2008)
Recast Nafion/(20 wt%) ZrSPP*	0.05 S cm^{-1} at 110°C and 98% RH	Kim et al. (2006)
Recast Nafion*/SGO	0.047 S cm^{-1} at 130°C and 30% RH	Zarrin et al. (2011)
SPEEK-WC	0.00022 S cm^{-1} at 96°C and 85% RH	Fontananova et al. (2010)
SPEEK-WC/SiW (9.6 wt%)	0.0013 S cm^{-1} at 96°C and 85% RH	Fontananova et al. (2010)
SPEEK/poly(imi-alt-CTFE)	0.001−0.015 S cm^{-1} at 120°C and 25−95% RH	Frutsaert et al. (2011)
SPEEK/ZrP-NS)	0.079 S cm^{-1} at 150°C and 100% RH	Kozawa et al. (2010)
SPI with fluorene groups	1.67 S cm^{-1} at 120°C and 100% RH	Miyatake et al. (2011)
SPI with fluorene groups	0.003S cm^{-1} at 160°C and 12% RH	Ye et al. (2006)
SPI with sulfophenoxypropoxy pendants	1 S cm^{-1} at 120°C and 100% RH	Miyatake et al. (2007)
SPU with fluorene	0.5 S cm^{-1} at 110°C and 50% RH	Bae et al. (2010)
CF$_6$-PBI	0.12 S cm^{-1} at 175°C and 10% RH	Li et al. (2008)
SO$_2$-PBI	0.12 S cm^{-1} at 180°C and 5% RH	Li et al. (2010)
PBI/SPAES	0.045 S cm^{-1} at 200°C and 0% RH	Lee et al. (2008)
PSU with pyridine and hydroquinone groups	0.02 S cm^{-1} at 120°C and 0% RH	Kallitsis et al. (2009)
72TEOS-18PDMS-10PO (OCH$_3$)$_3$/40% [EMI][TFSI]	0.00487 S cm^{-1} at 150°C and 0% RH	Lakshminarayana et al. (2010)
TMOS-TMPS-TEP/(40%) BMIMBF$_4$	0.006 S cm^{-1} at 150°C and 0% RH	Lakshminarayana et al. (2011a,b)
30TMOS-30TEOS-30MTEOS-10PO (OCH$_3$)$_3$/(40%) EMIMBF$_4$	0.01 S cm^{-1} at 155°C and 0% RH	Lakshminarayana et al. (2011a,b)
60TMOS-30VTMOS-10PO(OCH$_3$)$_3$/(40%) EMIMBF$_4$	0.0089 S cm^{-1} at 155°C and 0% RH	Lakshminarayana et al. (2011a,b)

*Membrane cast from 5 wt% Nafion 1100 solution, DuPont.
**Membrane cast from Nafion resin, Shandong Dongyue Polymer Material Co. Ltd.

Table 3.2 Structures of the Most Common HT Membrane Materials

Name	Structure	References
Nafion/PFSA		Banerjee and Curtin (2004)
PBI		Asensio et al. (2010)
SPEEK		Iojoiu et al. (2006)
SPI		Miyatake et al. (2011)
SPSU		Parvole and Jannasch (2008)

Table 3.3 HT-MEA Preparation Methods (Chandan et al., 2013)

Membrane Material	GDL	Catalyst	Application	Hot Press
PBI doped with phosphotungstic acid and silicotungstic acid	Carbon cloth (Etec Inc., West Peabody, Massachusetts, United States)	Catalyst ink: Platinum on Vulcan (XC-72; Etec Inc., 20 wt% anode, 40 wt% cathode), PBI binder (5 wt% in DMAc), DMAc solvent. PT loading: anode 0.2 mg cm^{-2}, cathode 0.4 mg cm^{-2}	Airbrush application onto GDL. Dried/sintered at 190°C overnight	GDLs onto membrane. 110°C, 0.4 ton cm^{-2}, 10min
SPESK block copolymer with sulfonated fluorene groups	Carbon paper (25BCH – SGL)	Catalyst ink: Platinum on carbon (TEC10E70TPM, Tanaka, 68 wt%), SPESK binder (SPESK/C ratio 0.7 wt%), DMAc solvent. Pt loading: 0.45 mg cm^{-2}	Pulse swirl-spray application onto GDL at 60°C. Washed with hot water and 1 M nitric acid at 60°C. Dried at 60°C under vacuum	GDLs onto membrane. No hot pressing

(Continued)

Table 3.3 (Continued)				
Membrane Material	GDL	Catalyst	Application	Hot Press
Ternary blend of sulfonated partially fluorinated arylene poly ether/PBI/H_3PO_4	Nonwoven, precoated with a microporous layer (type not given)	Catalyst ink: Platinum on carbon black (HiSPECTM 8000; Johnson Matthey, London, UK) Platinum catalysts, 48.6 wt%), PBI binder (Pt/PBI ratio 13), formic acid and phosphoric acid solvent. Pt loading: 0.6–0.7 mg cm^{-2}.	Spray application onto GDL. Dried at 80°C for an hour	GDLs onto membrane. 200°C, 0.1 ton cm^{-2}, 10 min
Recast Nafion doped with sulfonated graphene oxide	No GDL—decal method used	Catalyst ink: Platinum on carbon (E-TEK, 20 wt%). 5% Nafion binder, glycerol solvent. Pt loading: anode 0.1 mg cm^{-2}, cathode 0.2 mg cm^{-2}	Catalyst ink painted onto decal Teflon blanks. Dried in oven. Repeated until desired catalyst loading achieved	Decals onto membrane 210°C, 110lbs cm − 2, 5min

Other methods include automatic catalyst spraying under irradiation (ACSUI), ultrasonic spraying (US), electrophoretic deposition (EPD), and hand spraying (HS) (Su et al., 2014a). The researchers found that superior CL structures can be obtained with electrodes prepared by ACSUI and US methods. Although the electrodes prepared by EPD method exhibited fine CL structures and high macropore volumes, the uneven distribution of the binder and catalyst particles in the CL decreased Pt utilization, resulting in low electrode activity. High fuel cell performances were obtained by the electrodes prepared with ACSUI and US techniques. Under normal operating conditions, the peak power densities reached $\sim +0.4 \text{ W cm}^{-2}$, and the current densities at $+0.6 \text{ V}$ were up to 0.24 A cm^{-2}. Electrochemical analysis revealed that the electrodes have high CL activities, low charge transfer resistances and high ECSA (electrochemical surface areas). Short-term durability tests showed good stability of the electrodes. Therefore, for HT-MEA fabrication, spraying methods (ACSUI and US) could be selected as the most suitable techniques of CL deposition.

3.3 DEVELOPMENT OF HT-MEA WITH IMPROVED Pt UTILIZATION

The improvement of Pt utilization for PEMFC has always been a hot topic in this field. For HT-PEMFC, the Pt loading used is normally over 0.7 mg cm^{-2}, so improving Pt utilization is especially meaningful to "real" applications of this type of fuel cell. Three main approaches have been reported for this purpose.

3.3.1 HT-MEA With Low Pt Loading

Lowering the Pt loading is always the first objective that many researchers attempt to achieve. For example, Su et al. (2014b) reported the use of an ultrasonic spray system for producing low Pt loadings HT-MEAs using the CCS fabrication technique, as shown in Fig. 3.1.

Their experiments showed that the ultrasonic spray coating method was very suitable for achieving high-performance values at low and ultralow Pt-loading MEAs: the Pt loading was as low as 0.35 mg cm^{-2}, and yielded satisfactory performances owing to the combination of improved reaction kinetics and mass transport. They compared the data with those obtained in the literature under similar operating conditions and showed that the performances of their GDEs were comparable to the commercial ones.

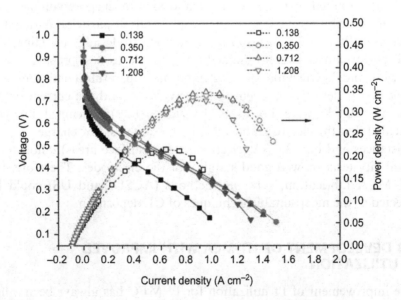

Figure 3.1 Polarization curves of the GDEs with different Pt loadings (Su et al., 2014b).

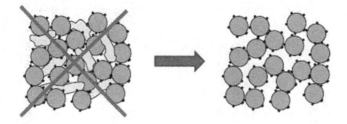

Figure 3.2 The schematic diagram of the new electrode concept for HT-PEMFCs without polymeric binder (Martin et al., 2014).

3.3.2 HT-MEA With Binderless Electrodes

Martin et al. (2014) proposed a new electrode concept for acid-doped PBI-based HT-PEMFCs in which no polymeric binder was used in the CL (Fig. 3.2).

They showed that a stable interface between the membrane and the CL can be retained when a proton-conducting acid phase is established. The absence of the polymer in the catalytic layer turned out to be beneficial for the PBI cell performance, particularly under high-load operation. The influence on performance of the Pt loading of the cathode was studied in the range from 0.11 to 2.04 mg Pt cm^{-2}, showing saturation of the maximum performance level for Pt loadings higher than 0.5 mg Pt cm^{-2}. For fuel cell operation on hydrogen and air supplied under ambient pressure, a peak power density as high as +471 mW cm^{-2} was obtained.

3.3.3 HT-MEA Prepared by the CCM Method

Another way researchers have been attempting to improve Pt utilization is by developing MEA based on the CCM method. As previously stated, the CCM method is widely used for the preparation of MEA for LT-PEMFCs based on Nafion membranes, because it can greatly reduce the Pt loading, compared to the CCS method. However, it is not a straightforward method for preparing HT-MEAs based on PBI or ABPBI membranes, because of the additional PA doping process. Liang et al. (2014, 2015) studied three different ways to prepare CCMs, based on different PA doping processes: through GDL, through membrane, and through the whole CCM. Their results showed that only CCM prepared with the PA doping GDL method yielded uniform MEA structures; the other two methods caused catalyst loss or serious distortion of the membrane. They evaluated the fuel cell performance of the CCM-based MEAs and showed that the HT-MEAs prepared by this method exhibited high Pt utilization when Pt

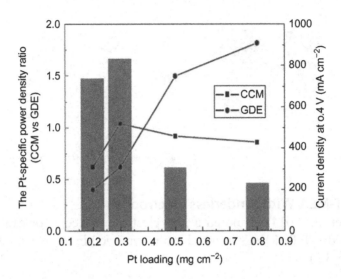

Figure 3.3 The Pt utilization ratio and the current density at +0.4 V comparison of MEAs prepared by CCM and GDE (Liang et al., 2014).

loading was lower than $0.3 \, \mathrm{mg_{Pt} \, cm^{-2}}$. When the Pt loading is increased, performance drops, while MEA prepared by the GDE method shows improved performance, as shown in Fig. 3.3.

It should be noted that the way of introducing PA for CCM-based MEAs is different from that for GDE-based MEAs. In GDE-based MEAs, the required PA for proton transfer relies upon PA diffusion from predoped PBI membrane, while the PA in the CCM-based MEAs comes from PA-impregnated GDLs. The PA must go through the whole CL before being absorbed by the membrane. The residual PA in the GDLs and the CLs may be higher than those for GDE-based MEAs, which could also cause difficulties in O_2 diffusion, resulting in performance deterioration, especially when the CL is thick. It is suggested that the diffusion of oxygen for the GDE-based MEA is preferable to that in the CCM-based MEA, as the risk of PA flood in the CLs should be much lower for the former. Therefore, the CCM method is a good technique when the Pt loading is relatively low, although it is not suitable if high MEA performance is required.

3.4 NOVEL CL STRUCTURE OF HT-MEA

An optimized CL structure can make a great improvement in the overall fuel cell performance. For instance, Su et al. (2014c,d) proposed a novel dual CL GDE for HT-MEAs, as shown in Fig. 3.4.

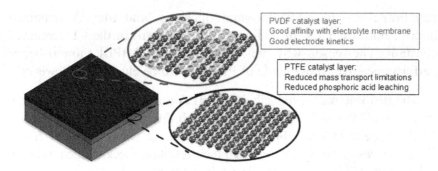

Figure 3.4 Conception of a novel dual CL GDE (Su et al., 2014c).

Differing from conventional GDEs with the CL-containing simplex binder, dual CL GDEs were prepared using two different binders, namely PVDF and PTFE. A PVDF CL was prepared as the outer layer in order to obtain good electrode kinetics by intimately contacting the electrolyte membrane, while a PTFE CL was prepared as the inner layer to reduce mass transport limitations. Single fuel cell tests and electrochemical analyses of both the dual-CL GDE and conventional GDEs were performed to evaluate the effect of the novel CL structure on the fuel cell performance. The results showed that significant reductions in both kinetics and mass transfer losses account for the enhanced performance of the novel dual-CL structured GDE.

3.5 DURABILITY AND REPRODUCIBILITY OF HT-MEA

Durability is an important parameter for "real" HT-PEMFC applications. Oono et al. (2013) reported enhanced lifetimes of a series of HT-MEAs using ABPBI membrane. Power generation experiments were carried out on these HT-MEAs with ABPBI membrane doped with PA to levels of 62%, 72%, and 76%, in order to evaluate the influence of the acid-doping level on fuel cell performance. Tests were also carried out on two cells with an acid-doping level of 78% at a current density of 0.2 A cm^{-2} for 1000 and 17,500 h in order to study the stability and durability of the electrolyte membrane. The long-term test results showed that the cell voltage decreases between $\sim 10\%$ and $\sim 4.4\%$ for the MEAs containing PBI membrane and ABPBI membrane, respectively (this was mainly due to accidental shutdowns of the test station during the testing period). The postanalysis data indicated that, following fuel cell operation for 17,500 h, the MEA made of ABPBI membrane maintained its original thickness, and the PA remained high in both the CLs. In addition, the Pt content in the CL remained constant. The reason why the MEAs made of ABPBI showed better performance than those made of PBI is yet to be elucidated, however.

The performance reproducibility of MEA is also an important parameter for "real" fuel cell applications, although there were almost no reports on this aspect. In the 2014 CARISMA Meeting, Su et al., from *HySA Systems Competence Centre* reported their result on checking the reproducibility of the HT-MEA they made, as shown in Fig. 3.5.

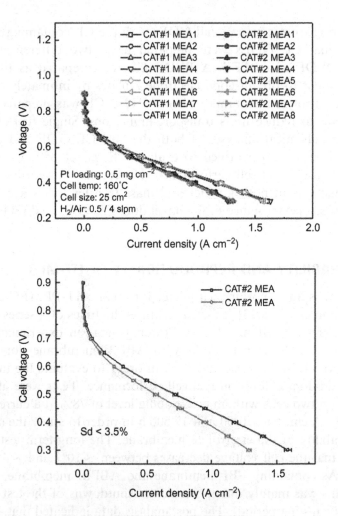

Figure 3.5 Performance reproducibility of the HT-MEAs made by HySA Systems.

They prepared eight MEAs separately by using two different catalysts, and then the performances of these MEAs were evaluated. The standard deviation of their performance was less than 3.5%, indicating good reproducibility of the produced HT-MEAs.

3.6 SUMMARY

The preparation of an optimized HT-MEA is an art and a science in itself. High fuel cell performance is highly dependent upon excellent MEA structures where the catalyst is properly dispersed, the support is stable and the ionomer optimally creates the 3-phase interface (TPB) without limiting mass transport. Prior to MEA preparation, it is necessary to understand how to structure the CL in such a way as to control the nanoparticle size, stability and wetting properties as well as its electronic properties; hence increasing the utilization of the catalyst. MEAs of high performance, good durability, and good reproducibility are now achievable, although further improvement is urgently needed.

REFERENCES

Asensio, J.A., Sanchez, E.M., Gomez-Romero, P., 2010. Proton-conducting membranes based on benzimidazole polymers for high-temperature PEM fuel cells. A chemical quest. Chem. Soc. Rev. 39, 3210–3239.

Bae, B., Yoda, T., Miyatake, K., Uchida, H., Watanabe, M., 2010. Proton-conductive aromatic ionomers containing highly sulfonated blocks for high-temperature-operable fuel cells. Angew. Chem. Int. Ed. 49, 317–320.

Banerjee, S., Curtin, D.E., 2004. Nafion® perfluorinated membranes in fuel cells. J. Fluorine Chem. 125 (8), 1211–1216.

Casciola, M., Capitani, D., Comite, A., Donnadio, A., Frittella, V., Pica, M., et al., 2008. Nafion-zirconium phosphate nanocomposite membranes with high filler loadings: conductivity and mechanical properties. Fuel Cells 8 (3–4), 217–224.

Chandan, A., Hattenberger, M., El-kharouf, A., Du, S., Dhir, A., Self, V., et al., 2013. High temperature (HT) polymer electrolyte membrane fuel cells (PEMFC)—a review. J. Power Sources 231, 264–278.

Fontananova, E., Trotta, F., Jansen, J.C., Drioli, E., 2010. Preparation and characterization of new non-fluorinated polymeric and composite membranes for PEMFCs. J. Membr. Sci. 348 (1–2), 326–336.

Frutsaert, G., David, G., Ameduri, B., Jones, D.J., Roziere, J., Glipa, X., 2011. Synthesis and characterisation of novel fluorinated polymers bearing pendant imidazole groups and blend membranes: new materials for PEMFC operating at low relative humidity. J. Membr. Sci. 367 (1–2), 127–133.

Iojoiu, C., Chabert, F., Maréchal, M., Kissi, N.E., Guindet, J., Sanchez, J.Y., 2006. From polymer chemistry to membrane elaboration: a global approach of fuel cell polymeric electrolytes. J. Power Sources 153 (2), 198–209.

Kallitsis, J.K., Geormezi, M., Neophytides, S.G., 2009. Polymer electrolyte membranes for high-temperature fuel cells based on aromatic polyethers bearing pyridine units. Polym. Int. 58 (11), 1226–1233.

Kim, Y.T., Kim, K.H., Song, M.K., Rhee, H.W., 2006. Nafion/ZrSPP composite membrane for high temperature operation of proton exchange membrane fuel cells. Curr. Appl. Phys. 6 (4), 612–615.

Kozawa, Y., Suzuki, S., Miyayama, M., Okumiya, T., Traversa, E., 2010. Proton conducting membranes composed of sulfonated poly(etheretherketone) and zirconium phosphate nanosheets for fuel cell applications. Solid State Ionics 181 (5–7), 348–353.

Lakshminarayana, G., Nogami, M., Kityk, I.V., 2010. Synthesis and characterization of anhydrous proton conducting inorganic-organic composite membranes for medium temperature proton exchange membrane fuel cells (PEMFCs). Energy 35 (12), 5260–5268.

Lakshminarayana, G., Vijayaraghavan, R., Nogami, M., Kityk, I.V., 2011a. Anhydrous proton conducting hybrid membrane electrolytes for high temperature (>100°C) proton exchange membrane fuel cells. J. Electrochem. Soc. 158 (4), B376–B383.

Lakshminarayana, G., Nogami, M., Kityk, I.V., 2011b. Novel hybrid proton exchange membrane electrolytes for medium temperature non-humidified fuel cells. J. Alloys Compd. 509 (5), 2238–2242.

Li, Q., Jensen, J.O., Pan, C., Bandur, V., Nilsson, M.S., Schonberger, F., et al., 2008. Partially fluorinated aarylene polyethers and their ternary blends with PBI and H_3PO_4. Part II. Characterisation and fuel cell tests of the ternary membranes. Fuel Cells 8 (3–4), 188–199.

Li, Q.F., Rudbeck, H.C., Chromik, A., Jensen, J.O., Pan, C., Steenberg, T., et al., 2010. Properties, degradation and high temperature fuel cell test of different types of PBI and PBI blend membranes. J. Membr. Sci. 347 (1–2), 260–270.

Liang, H., Su, H., Pollet, B.G., Linkov, V., Pasupathi, S., 2014. Membrane electrode assembly with enhanced platinum utilization for high temperature proton exchange membrane fuel cell prepared by catalyst coating membrane method. J. Power Sources 266, 107–113.

Liang, H., Su, H., Pollet, B.G., Pasupathi, S., 2015. Development of membrane electrode assembly for high temperature proton exchange membrane fuel cell by catalyst coating membrane method. J. Power Sources 288, 121–127.

Lee, H.S., Roy, A., Lane, O., McGrath, J.E., 2008. Synthesis and characterization of poly(arylene ether sulfone)-b-polybenzimidazole copolymers for high temperature low humidity proton exchange membrane fuel cells. Polymer 49 (25), 5387–5396.

Martin, S., Li, Q., Steenberg, T., Jensen, J.O., 2014. Binderless electrodes for high-temperature polymer electrolyte membrane fuel cells. J. Power Sources 272, 559–566.

Mazúr, P., Soukup, J., Paidar, M., Bouzek, K., 2011. Gas diffusion electrodes for high temperature PEM-type fuel cells: role of a polymer binder and method of the catalyst layer deposition. J. Appl. Electrochem. 41, 1013–1019.

Miyatake, K., Yasuda, T., Hirai, M., Nanasawa, M., Watanabe, M., 2007. Synthesis and properties of a polyimide containing pendant sulfophenoxypropoxy groups. J. Polym. Sci. A Polym. Chem. 45, 157–163.

Miyatake, K., Bae, B., Watanabe, M., 2011. Fluorene-containing cardo polymers as ion conductive membranes for fuel cells. Polym. Chem. 2, 1919–1929.

Nam, S.E., Kim, S.O., Kang, Y., Lee, J.W., Lee, K.H., 2008. Preparation of Nafion/sulfonated poly(phenylsilsesquioxane) nanocomposite as high temperature proton exchange membranes. J. Membr. Sci. 322 (2), 466–474.

Oono, Y., Sounai, A., Hori, M., 2013. Prolongation of lifetime of high temperature proton exchange membrane fuel cells. J. Power Sources 241, 87–93.

Park, J.O., Kwon, K., Cho, M.D., Hong, S.G., Kim, T.Y., Yoo, D.Y., 2011. Role of binders in high temperature PEMFC electrode. J. Electrochem. Soc. 158, B675–B681.

Parvole, J., Jannasch, P., 2008. Polysulfones grafted with poly(vinylphosphonic acid) for highly proton conducting fuel cell membranes in the hydrated and nominally dry state. Macromolecules 41 (11), 3893–3903.

Su, H., Pasupathi, S., Bladergroen, B., Linkov, V., Pollet, B.G., 2013. Optimization of gas diffusion electrode for polybenzimidazole-based high temperature proton exchange membrane fuel cell: evaluation of polymer binders in catalyst layer. Int. J. Hydrogen Energy 38, 11370–11378.

Su, H., Felix, C., Barron, O., Bujlo, P., Bladergroen, B., Pollet, B., et al., 2014a. High-performance and durable membrane electrode assemblies for high-temperature polymer electrolyte membrane fuel cells. Electrocatalysis 5, 361–371.

Su, H., Jao, T.-C., Barron, O., Pollet, B.G., Pasupathi, S., 2014b. Low platinum loading for high temperature proton exchange membrane fuel cell developed by ultrasonic spray coating technique. J. Power Sources 267, 155–159.

Su, H., Jao, T.-C., Pasupathi, S., Bladergroen, B.J., Linkov, V., Pollet, B.G., 2014c. A novel dual catalyst layer structured gas diffusion electrode for enhanced performance of high temperature proton exchange membrane fuel cell. J. Power Sources 246, 63–67.

Su, H., Liang, H., Bladergroen, B.J., Linkov, V., Pollet, B.G., Pasupathi, S., 2014d. Effect of platinum distribution in dual catalyst layer structured gas diffusion electrode on the performance of high temperature PEMFC. J. Electrochem. Soc. 161, F506–F512.

Verma, A., Scott, K., 2010. Development of high-temperature PEMFC based on heteropolyacids and polybenzimidazole. J. Solid State Electrochem 14, 213–219.

Ye, X.H., Bai, H., Ho, W.S.W., 2006. Synthesis and characterization of new sulfonated polyimides as proton-exchange membranes for fuel cells. J. Membr. Sci. 279 (1–2), 570–577.

Yuan, J.J., Pu, H.T., Yang, Z.L., 2009. Studies on sulfonic acid functionalized hollow silica spheres/Nafion® composite proton exchange membranes. J. Polym. Sci. A Polym. Chem. 47, 2647–2655.

Zarrin, H., Jun, Y., Fowler, M., Chen, Z., 2011. Functionalized graphene oxide as a new highly proton conductive composite membrane for high temperature PEMFCs. In: 219th ECS Meeting©, Abstract #643. The Electrochemical Society.

CHAPTER 4

HT-PEMFC Modeling and Design

4.1 INTRODUCTION

A high-temperature polymer electrolyte membrane fuel cell (HT-PEMFC) stack includes serially connected single unit cells to produce the designed output power, as shown schematically in Fig. 4.1. Each cell consists of a membrane for ion conduction, two catalyst layers for electro-chemical reactions, two gas diffusion layers, and two bipolar plates for electron conduction and flow distribution. The gas diffusion layers and the bipolar plates are the same as those used in low-temperature polymer electrolyte membrane fuel cells (LT-PEMFCs). Fuel cells have several length scales of significance: of the order of nanometers at the platinum catalyst level at which the charge (electron) transfer reactions take place; of the order of micrometers at which charge and reactant transports take place through the catalyst layers and the gas diffusion layers; of the order of millimeters at which transport of uncharged chemical species constituting the reactants and products takes place; and of the order of centimeters at which heat transfer and fluid flow effects are important. It is not possible to have a single model which encompasses an exact resolution of all the levels and multiscale modeling is often used, as is evident from the literature review.

4.2 MODELING AND SIMULATION OF HT-PEMFC

Modeling and simulation of all components in HT-PEMFCs are important tools in providing additional understanding of fuel cell operational behavior. The use of mathematical models is one possible way to analyze species concentrations, temperature gradients, and pressure distributions within the fuel cell; these are very important parameters for predicting performance and durability of the overall HT-PEMFC in varying operating conditions. These are either steady-state or dynamic models using isothermal or nonisothermal conditions

Recent Advances in High-Temperature PEM Fuel Cells. DOI: http://dx.doi.org/10.1016/B978-0-12-809989-6.00004-9
© 2016 Elsevier Ltd. All rights reserved.

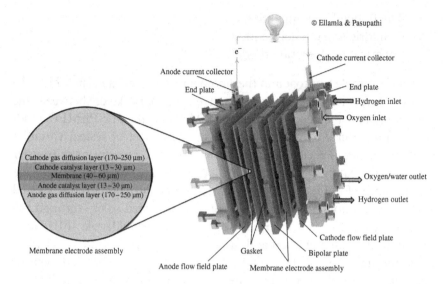

Figure 4.1 HT-PEMFC stack.

based on a single cell or stack level. Catalyst layers are concerned with in situ electrochemical modeling of phenomena occurring in parts of a single cell and these models are more theoretical and more single-cell design oriented. Another type of model is the control-oriented dynamic model based on the external load characteristics at single cell or stack level. A recent literature survey of HT-PEMFC modeling is listed in the following sections.

4.2.1 Catalyst Layer Models

A limited number of HT-PEMFC catalyst layer models is available in literature. The PEM fuel cell catalyst layer can be modeled via different approaches; there are two principal approaches, namely, macroscopic models that consider the catalyst layer as a whole, and microscopic models that consider the transport phenomena at the pore level.

The macroscopic model can be further classified into three categories.

Macroscopic models (Rao et al., 2007; Srinivasrao et al., 2010):

- Homogeneous (porous/nonporous)
- Film model
- Agglomerate models

- The flooded agglomerate model
- Cylindrical agglomerate model
- Spherical agglomerate model

A very simple approach is the homogeneous approach, in which the catalyst layer is treated as a reactive boundary layer between the membrane and the gas diffusion layer. Many of the HT-PEMFC models are limited in use and fail to show the real effect of system variables on performance. This approach does not take into account mass transfer effects in the catalyst layer, and this has led to an overestimation of the performance of the fuel cell, particularly at high current densities, caused by the assumption that mass transport has occurred solely through the porous media. In reality, an electrolyte (polybenzimidazole (PBI)/acid) thin film surrounding the catalyst particles is present and mass transport through this phase should be considered. In the second type of approach the catalyst layer is treated as a thin film of electrolyte flooded with a liquid water/acid layer. The film of electrolyte covering the catalyst particles is shown in Fig. 4.2A. In this film, reactants have to dissolve in the electrolyte media and diffuse through it to reach the catalytic sites (Siegel et al., 2011). Oxygen permeability through the thin electrolyte film varies based upon temperature, current density and the equilibrium vapor pressure of the product water above the thin film. Relatively thick catalyst layers with very low porosity have been used in attempts to compensate for not modeling the thin film explicitly and to try and match the experimental data (Cheddie and Munroe, 2006e; Mamlouk et al., 2011). Most models used a reaction order in the range of 0.5−2 and varied the transfer coefficient (α) according to the doping level of the acid. The third approach is an agglomerate model, in which the catalyst layer is considered as agglomerate with porous interagglomerates of spaces filled with a mixture of electrolyte, reactant, and products. According to the agglomerate model, carbon supported catalyst particles flood within the electrolyte and form agglomerates covered with a thin film of electrolyte, as shown in Fig. 4.2B. The agglomerates are of various shapes, such as spherical or cylindrical. The catalyst layer consists of macro−microporous, interconnected, hydrophobic regions so as to allow the reactant gas to access the surface of the agglomerate regions. The flooded agglomerate model is very successful in explaining oxygen diffusion in hydrophobic pores and electrolyte thin films. Sun et al. (2005) developed a spherical agglomerate model for an LT-PEMFC anode catalyst layer, and Karan (2007) developed a spherical

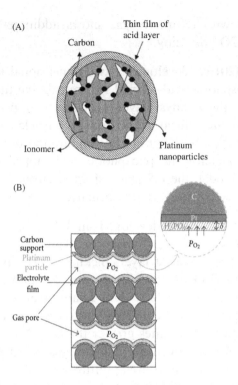

Figure 4.2 Schematics of catalyst layer using (A) the thin film model (Mamlouk et al., 2011) and (B) spherical agglomerate model.

agglomerate model for an LT-PEMFC cathode catalyst layer (CCL). The agglomerate model gives a better representation of the catalyst layer, judging on the basis of the simulations and a structural study of the catalyst layer (Rao et al., 2007). Many researchers have suggested that spherical agglomerate models are a more appropriate choice for predicting mass transfer losses, even at a low cell voltage of operation.

Mamlouk et al. (2011) developed a one-dimensional (1D) model of a PBI-membrane HT-PEMFC. This model addressed mass transport through a thin film electrolyte covering the catalyst particles, as well as through porous media, as shown in Fig. 4.2A. The catalyst interface is represented by a macro-homogeneous model. The model also covered the influence of the reformate gas (carbon monoxide, carbon dioxide, and methane) in terms of its effect on the anode polarization/kinetics behavior. It gave good predictions of the effect of oxygen and air pressures on cell behavior and also mass transport behavior within the cell. The model with reformate gas shows additional voltage losses associated with CO poisoning.

Sousa et al. (2010a) developed a two-dimensional (2D) isothermal model for a phosphoric acid−doped PBI membrane fuel cell; to study the influence of the catalyst layer properties on performance. The results show that the utilization of catalyst particles was very low at high current densities. Sousa et al. (2010b) developed a 2D nonisothermal dynamic model for the phosphoric acid−doped PBI membrane fuel cell. In this model, the influence of an electrode double layer was investigated with step change in performance.

Srinivasrao et al. (2010) developed an LT-PEMFC catalyst layer model with thickness varying from 10 to 30 μm. The model predicted the important role of catalyst layer thickness in cell performance and concluded that the ORR depends upon rate control mechanisms in the LT-PEMFC. Zhang et al. (2007) and Jalani et al. (2006) found that the mass transfer losses due to gas diffusion are significant in an HT-PEMFC at current densities greater than $0.4 \, A \, cm^{-2}$. An increase in cell temperature can effectively increase gas diffusivity but reduces solubility (or gas concentration in diffusion media). When the cell temperature increases, the effect of reduced gas solubility might be larger than the effect of increased gas diffusion resistance. As cell temperature increases, charge transfer and proton transfer resistances reduce. The activation energy of proton conductivity is almost independent of current density, indicating that the fuel cell operating load has no effect on the proton-conducting mechanism in the phosphoric acid−doped PBI membrane.

Based on the earlier LT-PEMFC catalyst layer models, Siegel et al. (2011) developed a three-dimensional (3D), steady-state, nonisothermal model for a phosphoric acid−doped PBI/H_3PO_4 sol−gel membrane of HT-PEMFC. Electrochemical reactions were modeled using an agglomerate approach and the effects of gas diffusivity and gas solubility were included. The conductivity of the membrane was modeled using the *Arrhenius* equation to describe its temperature dependence. It was observed that the *Arrhenius* approach was valid in a certain temperature range (150−160°C) and that the model overpredicted the PBI/H_3PO_4 sol−gel membrane conductivity at higher solid-phase temperatures.

4.2.2 Fuel Cell Models

Fuel cell models can be categorized as analytical, semiempirical, or mechanistic. Mechanistic models can be further subcategorized based upon the solution strategy, whether single- or multidomain (Cheddie and Munroe, 2005). Analytical models may be useful if rapid calculations are required for simple systems. Semiempirical modeling combines theoretically derived differential and algebraic equations with empirically determined relationships. Empirical relationships are employed when the physical phenomena are difficult to model or the theory governing the phenomena is not well understood. Semiempirical models are, however, limited to a narrow range of operating conditions. They cannot accurately predict performance outside of that range. They are very useful for yielding fast predictions for designs that already exist. Mechanistic modeling has received the most attention in the literature. In mechanistic modeling, differential and algebraic equations are derived based upon the physics and electrochemistry governing the phenomena internal to the cell. These equations are solved using some sort of computational method. Mechanistic modeling (single- and multidomain) has been employed to study a wide range of phenomena including polarization effects (activation, ohmic, and concentration overpotentials), water management, thermal management, CO kinetics, catalyst utilization, and flow field geometry (Bernardi, 1990; Bernardi and Verbrugge,1992). One-dimensional models are focused on fundamental electrochemical and transport phenomena in the gas diffusion layer, catalyst layer, and membrane. Two-dimensional numerical models are focused on gas flow channels, gas diffusion layers, and catalyst layers; and these models are solved using computational fluid dynamics (CFD). Three-dimensional numerical models focused on solving conservation equations of mass, momentum, and species were solved in the gas channels and gas diffusion media, while electrochemical kinetics were taken into account in source/sink terms (Song and Meng, 2013). The literature reviewing HT-PEMFC models is shown in Table 4.1.

In most fuel cell models, the membrane electrode assembly (MEA) is treated as a single layer and it is considered a reactive boundary between the anode and the cathode. In other models, the MEA is treated as two separate catalyst layers, and each layer is modeled in detail, as discussed in Section 4.2.1. Most of the HT-PEMFC models deal with the

Table 4.1 Current HT-PEMFC Models

Nonisothermal Model	Analytical/Semiempirical Model	Dynamic Model	1D	2D	3D	Year	Authors
	x		x			2006	Cheddie and Munroe (2006c)
	x		x			2011	Kulikovsky and McIntyre (2011)
				x		2006	Cheddie and Munroe (2006d)
				x		2007	Scott et al. (2007)
				x		2007	Cheddie and Munroe (2007)
				x		2010	Sharamardina et al. (2010, 2012)
				x		2010	Sousa et al. (2010a)
					x	2006	Cheddie and Munroe (2006b)
					x	2010	Lobato et al. (2010a,b)
					x	2011	Siegel et al. (2011)
					x	2011	Lobato et al. (2011)
					x	2012	Sousa et al. (2012)
x					x	2010	Jiao and Li (2010)
x					x	2009	Ubong et al. (2009)
	x					2006	Cheddie and Munroe (2006a)
	x					2008	Cheddie and Munroe (2008)
	x					2009	Scott and Mamlouk (2009)
	x					2010	Kulikovsky et al. (2010)
			x			2006	Hu et al. (2006)[a]
		x				2006	Jalani et al. (2006)[a]
		x				2010	Oono et al. (2010)[a]
		x		x		2010	Bergmann et al. (2010)[a,,b]
						2007	Zhang et al. (2007)[b]
				x		2006	Peng and Lee (2006)[b]
		x	x			2011	Mamlouk et al. (2011)[b]
		x			x	2008	Peng and Lee (2006)
x	x			x		2010	Sousa et al. (2010b)

[a]Degradation models.
[b]CO models.

electrochemical reactions at the electrodes using the *Butler–Volmer* or *Tafel* equations. Different types of HT-PEMFC models available in the literature are discussed in detail in this section.

Cheddie and Munroe (2006a) developed a 1D parametric model for HT-PEMFC, considering the effect of temperature and porous media characteristics on polarization performance. This parametric model failed to predict the effect of mass-transfer and could not explain limiting current (I_{lim}) observed under air operation. Later, they developed a 1D, single-phase, steady-state analytical model to predict the polarization performance of an HT-PEMFC (Cheddie and Munroe, 2006c). The *Tafel* approximation was used to describe the electrode kinetics and polarization curve fitted using least-squares with a reaction order equal to 0.5. This model showed a better fit with air than with oxygen at low current densities (i) but it underestimated performance at high current densities. The same authors later developed a 2D model, which predicts the temperature and concentration profiles in the fuel cell. This model accounts for rib effects and the variation of transport properties along the gas channels (Cheddie and Munroe, 2006d). This model does not take into account reactant gases dissolved in the catalyst layer, which is assumed to be macro-homogenous. The same authors later developed a 3D model with the assumptions (1) that the transfer coefficient (α) is equal to two and (2) that the reaction order is equal to one (Cheddie and Munroe, 2006b). The simulation data underestimated performance compared to the experimental data with oxygen and overestimated performance during air operation at high current densities. The same authors later developed a two-phase, 2D model by taking account of electrolyte conductivity and the solubility of hydrogen and oxygen as a function of temperature (Cheddie and Munroe, 2007). The model was also used to investigate the dependence of the fuel cell performance on membrane doping level, catalyst activity, and transport properties of dissolved gases in the electrolytic membrane. This model concluded that only 1% of the catalyst surface was used for the reaction and once again failed to predict the polarization curve in low current density regions.

Hu et al. (2006) developed 1D degradation models to simulate the steady-state polarization curves recorded at different times during the aging tests. The model showed that the main reason for performance degradation was the decrease of the electrochemical surface area, caused by catalyst agglomeration during the high-temperature sintering process. No obvious performance degradation was found on PBI-membrane. The models again failed to show any apparent mass transport limitations under air operation in a limiting current region.

Scott et al. (2007) proposed a 1D model for PBI membrane-based HT-PEMFC. The electrode kinetics was described by the *Butler–Volmer* equation and the mass transport by the multicomponent *Stefan–Maxwell* equations, coupled with *Darcy's Law*. The effect of partial pressure effect distribution on the cell voltage and the power density was also considered. The model had a good fit with the experimental data but failed to show limiting current behavior under air operation.

Cheddie and Munroe (2008) developed a semianalytical model based upon CFD by using volumetric catalyst source terms as the interfacial boundary condition on the MEA. These semianalytical solutions matched very well with a full computational model in terms of the polarization results, and hydrogen and oxygen concentrations. These results showed that using analytical techniques did not compromise the accuracy of the model.

Scott and Mamlouk (2009) developed a simple semiempirical zero-order model for estimating cell voltage and power performance as a function of current density. The model considered the influence of electrode kinetics using the *Butler–Volmer* equation, over the complete cell voltage range, ohmic potential losses, and the effect of mass transport through the electrolyte films covering the catalyst layers on kinetics and thermodynamics. The model failed to explain cell voltage curve at temperatures above 120°C under the high-current density region.

Shamardina et al. (2010) developed a simple and fast solvable *pseudo-steady-state* isothermal model, taking into account crossover effects. The crossover effects are observed only within a low stoichiometric region near the limiting current density. It is unable to explain the initial drop of cell voltage. Kulikovsky et al. (2010) developed an analytical HT-PEMFC model and discussed the importance of basic kinetic and transport models. This model is limited in addressing only transportation losses of oxygen in the CCL, neglecting the anode transportation losses and anode overpotential.

Oono et al. (2010) investigated the relationship between the HT-PEMFC operation temperature and cell durability in terms of deterioration mechanisms. They observed that the cell voltage increased by approximately +100 mV when the cell temperature was increased from 120°C to 140°C. The thermodynamic open circuit voltage decreases with increasing temperature, owing to the increase of water partial pressure. When cell temperature was subsequently increased from 140°C to 200°C in increments of 10°C, cell voltage increased at a rate of approximately +10 mV per 10°C. Higher cell temperatures were found to result in higher cell voltages, but decreased cell life. The ohmic resistance drops slightly with increasing cell temperature and the mass transport resistance also decreases slightly as the cell temperature increases. Charge transfer resistance is reduced with increasing cell temperature. This leads to an increase in catalytic activity (better electrode reactions) and in turn to an increase in cell voltage. The reduction in cell voltage of approximately +20 mV during the long-term tests was considered to be due to the aggregation of the electrode catalyst particles in the early stage of power generation, in addition to the effects of crossover due to the depletion of phosphoric acid in the terminal stage, which occurs regardless of the cell temperature. Moreover, Jalani et al. (2006) showed that the activation overpotential of the HT-PEMFC was reduced as cell temperature increased.

Sousa et al. (2010a) developed a 2D isothermal model, in which the electrochemical reactions were described using the spherical agglomerate model. The model was validated with experimental data and it showed that the utilization of the catalyst particle was very low at high current densities. The results from the model and the experimental results showed that an optimum performance fuel cell can be obtained when the volume fraction of phosphoric acid in the catalyst is in the range of 30−55%. The same authors used the finite element method (FEM) to solve the nonisothermal model such as to study the influence of two different geometries (along and across the channel direction) on the cell performance (Sousa et al., 2010b). The authors reported large temperature differences through the MEA, which could occur if the catalyst layer was not efficiently used. They performed dynamic simulations and investigated the influence of the electrode double layer when subjected to a step change in cell potential. A current overshoot was found when a step change was performed; this overshoot was caused by the delayed change of local oxygen concentration, lagging behind the change in potential. This overshoot could be removed by increasing the double-layer capacitance. The same model was modified in order to understand MEA degradation over time; it was found that during the first period of 300 h a loss of catalyst activity was observed owing to the change in mean particle/agglomerate size, which was the dominant effect compared to the phosphoric acid loss. The observation of the degradation mechanism was in very strong agreement with the earlier study by Hu et al. (2006), in which a 500-h aging test was performed.

Sousa et al. (2012) developed a 3D isothermal model for an HT-PEMFC equipped with phosphoric acid doped PBI and tested it for different flow field topologies. The results proved that interdigitated flow field topology gives the highest power output. However, it was not suitable for the fuel cell system because hot spots were generated owing to the heterogeneous current density distribution. In this study, a new geometry was suggested in order to homogenize the mass flux in straight channel geometry with varying inlet and outlet manifold.

The influence of CO poisoning at the anode of an HT-PEMFC was investigated by many researchers. The adsorption of CO on Pt is associated with high negative entropy, indicating that adsorption is disfavored at high temperatures. CO tolerance is dramatically enhanced, from 10 to 20 ppm at 80°C to 1000 ppm at 130°C, and up to 30,000 ppm at 200°C (Mamlouk et al., 2011; Zhang et al., 2007). Recent HT-PEMFC catalysts and membranes have CO tolerance of up to 50,000 ppm and it is possible to use directly hydrogen produced from a simple (and cost-effective) reformer.

Bergmann et al. (2010) developed a dynamic and nonisothermal 2D model of a PBI-based HT-PEMFC, the results of which were validated by in-house experimental data. The anode catalyst layer was taken as a thin film in between the membrane and the GDL. The temperature dependency of the fuel cell performance and CO poisoning of the anode was analyzed by generating polarization curves at different CO concentrations as well as CO pulses. The analysis showed nonlinear behavior of fuel cell performance under the influence of CO.

Peng and Lee (2006) presented a single-phase, 3D and nonisothermal numerical model which was implemented into a CFD code. The current density increased with increasing operating temperature and the maximum temperature was found to be on the catalyst layer. The model showed that the width and distribution of the gas channel and the current collector lands were key optimization parameters for fuel cell operation. This model was extended by Peng et al. (2008), who described the transient behavior of the current density of the cell. The prediction showed transience in cell current density which overshot (undershot) the stabilized state value when the cell voltage was abruptly decreased (increased). It was found that the peak of the overshoot was related to the cathode air stoichiometric rather than the anode hydrogen stoichiometric. The maximum temperature was located in the CCL and both the fuel cell average temperature and the temperature deviation were increased with increasing current load.

Jiao and Li (2010) developed a 3D and nonisothermal model to investigate the effects of operating temperature, phosphoric acid doping level of the PBI membrane, inlet relative humidity, stoichiometry ratios of the feed gases, operating pressure, and air/oxygen on the cell performance. The model shows that increasing both the operating temperature and the phosphoric acid doping level are favorable for improving the fuel cell performance. Humidifying the feed gases produces negligible improvement on the fuel cell performance. Using oxygen instead of air produces significant improvements on the fuel cell performance, and increasing the stoichiometric ratios helps prevent concentration losses only at high current densities.

Ubong et al. (2009) developed a nonisothermal and 3D model, and validated it with a single cell with a triple serpentine channel flow field. The results showed that there was no drastic decrease in the cell voltage at high current density, owing to mass-transfer limitations; it was also concluded that reactants did not need to be humidified.

Lobato et al. (2010a) developed a 3D fuel cell geometry model with a cell active area of 50 cm². Using this CFD model, they showed that the current density distribution was directly linked to the way reactants were spread over the electrode surface. The model predicted slightly higher limiting current densities when using serpentine geometries.

Siegel et al. (2011) developed a 3D and isothermal model, where the electrochemical kinetics, the agglomerate approach, and membrane conductivity were examined by *Arrhenius* equations and the overall model validated with a six-channel cell. Good agreement with experimental results was found in the temperature range of 150–160°C. An artificial neural network approach was successfully applied to predict the polarization curve for an HT-PEMFC (Lobato et al., 2010b). Tortuosity was used as a model parameter to describe the influence of polytetrafluoroethylene content in the GDL.

Kulikovsky et al. (2010) developed an analytical model which was based upon a two-step procedure to evaluate parameters such as exchange current density (i_o), *Tafel* slope (b), and cell resistance (R_{cell}) from two sets of polarization curves for an HT-PEMFC and validated it with experimental data. Shamardina et al. (2010) developed an analytical model, 2D pseudo, steady-state and isothermal model which accounted for the crossover of reactant gases through the polymeric membrane. The model results showed that the crossover effect had a considerable influence only at low temperatures. The same authors developed a model which took into account transport losses in the CCL (Shamardina et al., 2012). This model led to a more accurate valuation of the exchange current density and provided useful data on the porosity and the effective oxygen diffusivity in the CCL. Most of the above models have in common a set of fundamental parameters which could be compared in order to check consistency.

4.2.3 Fuel Cell Stack Models

A fuel cell stack consists of repeating units of single cells and cooling plates (with coolant flow field channels) in series. There are several approaches to constructing a stack level model, depending upon the modeling goal requirements. The HT-PEMFC stack modeling approaches are discussed in Table 4.2. Type (I): in which the full stack is described as a zero-dimensional model without any spatial resolution. These contain less detailed information at cell level, since the main focus on the stack overall performance is within a system. Type (II): in which the information of all the cells is averaged, i.e., one "effective" cell (or part of it) is modeled or a small portion of the stack (by taking the symmetry boundary condition), and the full stack is obtained by multiplying the number of cells in the stack. Type (III): in which the single cells are modeled explicitly and are coupled (e.g., by

Table 4.2 HT-PEMFC Stack Models			
Type of Approach	Description	Advantages	Authors
Zero-dimensional stack model	Modeled without any spatial resolution	Model contains little information about the cell level	Korsgaard et al. (2006a, b, 2008a,b); Chrenko et al. (2010); Ahluwalia et al. (2003)
Single cell or small portion of stack model	Information of all the cells is averaged	Reduces the computational time	Andreasen and Kær (2009); Reddy and Jayanti (2012); Reddy et al. (2013)
Explicitly single cell model	Single cells are coupled to yield the full behavior of the stack	It allows a narrow specific task	Chang et al. (2006); Cheng and Lin (2009); Hawkes et al. (2009)

the stack manifolds for flow) to yield the full behavior of the stack. It is often advantageous to narrow the specific task in a way that allows for a suitable model reduction.

In the first type of approach, the model contains little information at cell level as it is mainly focused on the overall performance of the stack. Korsgaard et al. (2006a,b) developed a semiempirical stack model and validated it with experimental data. The cell voltage was calculated as a function of temperature, current density, and air stoichiometry. The CO content in the anode varied from 0% to 5%, with CO_2 content ranging from 25% to 20% and remaining 75% H_2 content and temperatures ranging from 160°C to 200°C. The performance of this model showed excellent agreement with the experimental data and the simplicity and accuracy of the model made it ideal for system modeling and "real-time" applications. The experimental results on pure hydrogen were used as test data to estimate key parameters by using the least squares optimization algorithm. Korsgaard et al. (2008a,b) successfully applied this type of modeling approach based upon their earlier models (Korsgaard et al., 2006a,b) to find static system integration as well as dynamical control strategies for a fuel cell stack based on PBI membranes. For that purpose a CHP system was used. The CHP consisted of an HT-PEMFC, which was integrated with a steam-reforming reactor, a burner, a heat reservoir, and other auxiliary equipment. Chrenko et al. (2010) developed a static and dynamic model for a diesel fuel processor feeding the fuel cell stack. The model was validated with experimental data. The structure of the temperature and mass flow controls in the fuel processor and supply system was derived.

Ahluwalia et al. (2003) developed an HT-PEMFC stack model which was based upon the performance of a gasoline reformer and an LT-PEMFC stack. A parametric study was conducted on the LT-PEMFC stack operating at a temperature of 80°C and the performance data were compared with those of the HT-PEMFC stack operating in the temperature range of 150–200°C. The model concluded that the HT-PEMFC stacks had higher efficiencies than the LT-PEMFC stacks.

In the second approach, part of the fuel cell stack constitutes the computational domain. This type of model reduces the overall computational domain to a small unit and substantially reduces the simulation time. The information of all cells is averaged, based upon the computational domain, and the overall stack information can be obtained by multiplying these individual cells or a model domain with a number of repeating units. Most of the models that are available in the literature have a repeat unit of one cell and one channel (Tao et al., 2006). This method was successfully applied to describe a 20-cell LT-PEMFC stack model to find the effects of reactant flows, two-phase effects and temperature distribution on the fuel cell stack performance (Park and Choe, 2008).

Andreasen and Kær (2009) developed an impedance-based HT-PEMFC stack model which is able to predict the stack impedance at different temperature profiles of the fuel cell stack. Simple equivalent circuit models for each single fuel cell can be used to predict the HT-PEMFC stack impedance at various temperatures. The full stack impedance depends upon the impedance of each of the single cells of the stack. Designing such a model allows the prediction of fuel cell behavior in steady-state as well as in dynamic operation, with an advantage in controlling fuel cell systems. This model is very useful in fuel cell system performance prediction, where different electronic components introduce current harmonics. A small portion stack was modeled by taking advantage of the symmetry boundary condition which predicted local temperature variations in the 1-kw HT-PEMFC stack with a cathode air and with a separate liquid coolant circuit— this model confirmed the importance of coolant inlet temperature (Reddy and Jayanti, 2012; Reddy et al., 2013).

In the third approach, single cells are modeled explicitly and are coupled (e.g., by the stack manifolds for flow) to yield a full behavior of the stack. Models of this type require more computational time when the full geometry is modeled. This approach is then successfully applied and demonstrated in a short LT-PEMFC stack (Chang et al., 2006; Cheng and Lin, 2009; Hawkes et al., 2009). The volume averaging method was applied for HT-PEMFC stacks in which the effective cell was modeled in 3D (Kvesic et al., 2012a,b). These models predicted the local current distribution inside the stack with a deviation of approximately 10%. The HT-PEMFC stack models associated with coolant strategies are discussed in Section 4.2.3.

4.2.4 Thermal Management Models

The cooling of a fuel cell stack can be achieved in a number of ways; it may include active cooling (with air or liquid coolants), passive cooling (with fins or heat spreaders), evaporative cooling or cooling with phase change liquids, and cooling with a separate air flow (Zhang and Kandlikar, 2012). Passive cooling methods have the limitation that they can be used only for very small stacks. An active cooling method, in which the coolant fluid is pumped through cooling passages within the stack, is capable of greater heat removal from the stack. This concept has attracted much attention for the efficient cooling of LT-PEMFCs (Yu et al., 2009; Cozzolino et al., 2011; Baek et al., 2011; Asghari et al., 2011). Thermal management of HT-PEMFC stacks requires two considerations, namely, the initial heat-up of the stack to operating temperature and later its maintenance at a constant operating temperature. A number of stack thermal management studies have been reported in the literature. These include the study of: Andreasen et al. (2008) for an air-cooled HT-PEMFC stack designed for a hybrid electric vehicle; the studies of Scholta et al. (2008, 2009) for a 5-cell, air/liquid cooled stack and a 10-cell stack cooled using heat pipes; Song et al. (2011) for a natural circulation-driven water-cooling system; and Kvesic et al. (2012a,b) who developed a multiscale, 3D model of a stack containing one coolant channel in each bipolar plate. Kvesic et al. (2012a) showed that with preheating of the reactants and coolant, the cell temperature variation can be reduced to within the range of 3−6 K for a hydrogen-fed stack and to about 9−10 K for a stack fed with reformate gas (Kvesic et al., 2012b). This is in strong agreement with the findings of Scholta et al. (2009) who reported a cell temperature variation from center to edge of approximately 56 K for an inlet coolant temperature of 373 K and cell operating temperature of 433 K. Reddy et al. modeled a 1-kWe

HT-PEMFC stack with a cathode air and with a separate liquid coolant circuit (Reddy and Jayanti, 2012; Reddy et al., 2013) confirmed the importance of coolant inlet temperature. Most of the detailed studies reported hitherto have been performed for small stacks, typically of 1 kWe or less. There is an increasing interest in stacks in the kilowatt range, as evidenced by the recent studies by Janßen et al. (2013) and Samsun et al. (2014). Samsun et al. (2014) studied a 5-kWe stack designed as part of an auxiliary power unit using an onboard reformer fueled by diesel and kerosene.

4.3 STACK DESIGN PRINCIPLES

As previously described, the main components of the fuel cell stack are membrane, porous diffusion layers, catalyst layers, gaskets, flow field channels, current collector plates with electrical connections, and end plates with fluid connections. The whole stack is assembled by the use of tie-rods and bolts.

4.3.1 Flow Field Design

Fuel cell performance is very sensitive to the flow rate of the reactants, so it is necessary that each cell in a stack receives approximately the same amount of the reactants. The performance of a fuel cell is sensitive to the flow rate of the reactants and each cell active area in a stack has to receive a uniform amount of reactant gases. The flow field may be square, rectangular, circular, hexagonal, octagonal, or irregular, the most common shapes being square or rectangular. The flow field orientation may be either vertical or horizontal. The orientation of the flow field may have some effect on the liquid water removal during the shutdown of the stack. Many researchers have shown that a serpentine channel flow field is a good choice for HT-PEMFC stacks.

4.3.2 Manifold Design

Fuel cell stacks with a large number of cells require a manifold with a uniform flow distribution to each cell. The manifolds feed the reactant gases to the active cell areas and also collect the unused gases and products of the reactions. The configuration of the gas flow manifolds for a fuel cell stack is therefore an important engineering problem where once again a balance needs to be struck between uniformity of flow distribution and minimizing pressure drops. Planar fuel cells have two types of manifolds, namely, internal and external manifolds.

External manifolds are simpler and less costly than the internal manifolds but have major problems with leakage and sealing. Internal manifolds have the advantage of better sealing but are more expensive and add weight to the stack. Given that HT-PEMFCs deal mostly with gaseous reactants and products (water is produced in vapor form), leakage from liquid water accumulation and corrosion is less of a problem. Most of the stacks are configured in either a "U" (where the inlet and outlet are on the same side) or a "Z" shape (where the inlet and outlet are opposite to each other) based on the requirement.

4.3.3 Bipolar Plates for an HT-PEMFC

Bipolar plates separate the reactant gases and distribute them on each side over the whole active area of the MEA. Bipolar plates also remove the unreacted gases and water from the active area of the MEA. Bipolar plates should be electrically conductive, highly chemically resistant to the operating conditions, and highly thermally conductive for better heat transfer across the cell. Bipolar plates for LT- and HT-PEMFCs are made of almost the same materials, but HT-PEMFC bipolar plate material has to endure a steady electrical potential, a low pH environment and temperatures of up to 200°C. It is required that the bipolar plates are electrically and thermally conductive. The latter requirement is another important feature that plays a pivotal role in temperature distribution across each cell and over the entire active area of a fuel cell stack. Bipolar plates can be manufactured in different ways, such as hot pressing and injection molding (Hamilton and Pollet, 2010). They can be made up of different materials, e.g., thermoplastics, metal, graphite, and a few other additives. The main benefit of injection molding is the production of such plates in very short cycle times, driving the cost down. Zentrum für BrennstoffzellenTechnik (ZBT) have successfully produced and demonstrated prototyped hot pressed bipolar plates for HT-PEMFC stacks (Brokamp, 2016). ZBT have also produced injection-molded bipolar plates of HT-PEMFC based on polyphenylene sulfide and showed that injection molding provides a very fast and cost-effective method.

4.3.4 Heat Removal From Stack

The cooling of a fuel cell stack can be achieved by either active or passive cooling. Owing to the wide range of power requirements in fuel cell applications, the fuel cell stack must be carefully sized and

Figure 4.3 (A) Fuel cell stack (40 cells) with an internally oil-cooled system (DTU stack). (B) Fuel cell stack (36) with an internally air- and liquid-cooled system (ZBT stack). (C) Fuel cell stack (30 cells) with an externally air-cooled system (McConnell, 2009). (D) Fuel cell stack (10 cells) with an externally oil-cooled system (Scholta et al., 2008; Bujlo et al., 2013).

various cooling options carefully chosen. Many researchers across the world have developed different cooling approaches for the HT-PEMFC stack, as shown in Fig. 4.3.

For example, the Technical University of Denmark (DTU) developed a 40-cell HT-PEMFC using an internal liquid-cooled system (Jensen et al., 2007), ZBT developed a 36-cell HT-PEMFC with an internal air/liquid-cooled system, generating 1 kWe electrical energy (Beckhaus, 2016), *Serenergy* developed an externally air-cooled system (McConnell, 2009) and Scholta et al. (2008) developed an external oil-cooling system. Generally, an external liquid coolant is a better option for thermal management of fuel cell stacks larger than 5 kWe, while an integrated air-cooling system can be used for fuel cell stacks ≤1 kWe.

4.4 SUMMARY

In this section, the detailed modeling studies of HT-PEMFC are discussed from catalyst layer to stack levels. The principles of fuel cell stack design and modeling for LT- and HT-PEMFC are essentially the same, although the selection of materials differs from low-temperature to high-temperature operations. Compared to LT-PEMFC, the thermal management of an HT-PEMFC stack is much simpler, but the heat rejection temperature difference between the coolant and the stack is complicated and thus the design of the coolant for the stack must be carefully selected. Transient heat transfer problems, as at the start-up and shut-down, need to be accurately modeled for developing appropriate control algorithms, especially in the case of automotive applications (with highly dynamic load profiles). HT-PEMFC stack model validation is still far from being mature, owing mainly to the lack of such facilities as testing protocols and test station designs.

REFERENCES

Ahluwalia, R.K., Doss, E.D., Kumar, R., 2003. Performance of high-temperature polymer electrolyte fuel cell systems. J. Power Sources 117, 45−60.

Andreasen, S.J., Kær, S.K., 2009. Dynamic model of the high temperature proton exchange membrane fuel cell stack temperature. J. Fuel Cell Sci. Technol. 6, 1−8.

Andreasen, S.J., Ashworth, L., Reman, I.N.M., Rasmussen, P.L., Nielsen, M.P., 2008. Modeling and implementation of a 1 kW, air cooled HTPEM fuel cell in a hybrid electrical vehicle. ECS Trans. 12, 639−650.

Asghari, S., Akhgar, H., Imani, B.F., 2011. Design of thermal management subsystem for a 5kW polymer electrolyte membrane fuel cell system. J. Power Sources 196, 3141−3148.

Baek, S.M., Yu, S.H., Nam, J.H., Kim, C.J., 2011. A numerical study on uniform cooling of large-scale PEMFCs with different coolant flow field designs. Appl. Therm. Eng. 31, 1427−1434.

Beckhaus I.P., ZTB stack: HT-PEM fuel cell stacks. <http://www.zbt-duisburg.de/en/portfolio/fuel-cell-stacks/ht-pem-stacks/> (last accessed May 2016).

Bergmann, A., Gerteisen, D., Kurz, T., 2010. Modeling of CO poisoning and its dynamics in HTPEM fuel cells. Fuel Cells 10, 278−287.

Bernardi, D.M., 1990. Water balance calculations for solid polymer electrolyte fuel cells. J. Electrochem. Soc. 137, 3344−3350.

Bernardi, D.M., Verbrugge, M.W., 1992. A mathematical model of the solid-polymer-electrolyte fuel cell. J. Electrochem. Soc. 139, 2477−2491.

Brokamp D.S., 2016. Development of bipolar plates for HT-PEM fuel cells. <http://www.zbt-duisburg.de/en/portfolio/fuel-cell-components/ht-bipolar-plates/> (last accessed May 2016).

Bujlo, P., Pasupathi, S., Ulleberg, Ø., Scholta, J., Nomnqa, M.V., Rabiu, A., et al., 2013. Validation of an externally oil-cooled 1 kWe HT-PEMFC stack operating at various experimental conditions. Int. J. Hydrogen Energy 38, 9847−9855.

Chang, P.A.C., St-Pierre, J., Stumper, J., Wetton, B., 2006. Flow distribution in proton exchange membrane fuel cell stacks. J. Power Sources 162, 340−355.

Cheddie, D.F., Munroe, N.D.H., 2005. Review and comparison of approaches to proton exchange membrane fuel cell modeling. J. Power Sources 147, 72−84.

Cheddie, D.F., Munroe, N.D.H., 2006a. Two dimensional phenomena in intermediate temperature PEMFCs. Int. J. Transport Phenomena 8, 51–68.

Cheddie, D.F., Munroe, N.D.H., 2006b. Three dimensional modeling of high temperature PEM fuel cells. J. Power Sources 160, 215–223.

Cheddie, D.F., Munroe, N.D.H., 2006c. Parametric model of an analytical correlations for intermediate temperature PEMFC.PEM fuel cells. J. Power Sources 160, 299–304.

Cheddie, D.F., Munroe, N.D.H., 2006d. Two dimensional phenomena in intermediate temperature PEMFCs. Int. J. Transp. Phenom. 8, 51–68.

Cheddie, D.F., Munroe, N.D.H., 2006e. Mathematical model of a PEMFC using a PBI membrane. Energy Convers. Manage. 47, 1490–1504.

Cheddie, D.F., Munroe, N.D.H., 2007. A two-phase model of an intermediate temperature PEM fuel cell. Int. J. Hydrogen Energy 32, 832–841.

Cheddie, D.F., Munroe, N.D.H., 2008. Semi-analytical proton exchange membrane fuel cell modeling. J. Power Sources 183, 164–173.

Cheng, C.H., Lin, H.H., 2009. Numerical analysis of effects of flow channel size on reactant transport in a proton exchange membrane fuel cell stack. J. Power Sources 194, 349–359.

Chrenko, D., Lecoq, S., Herail, E., Hissel, D., Péra, M.C., 2010. Static and dynamic modeling of a diesel fed fuel cell power supply. Int. J. Hydrogen Energy 35, 1377–1389.

Cozzolino, R., Cicconardi, S.P., Galloni, E., Minutillo, M., Perna, A., 2011. Theoretical and experimental investigations on thermal management of a PEMFC stack. Int. J. Hydrogen Energy 36, 8030–8037.

Hamilton, P.J., Pollet, B.G., 2010. Polymer electrolyte membrane fuel cell (PEMFC) flow field plate: design, materials and characterisation. Fuel Cells 10 (4), 489–509.

Hawkes, G., Obrien, J., Stoots, C., Hawkes, B., 2009. 3D CFD model of a multi-cell high-temperature electrolysis stack. Int. J. Hydrogen Energy 34, 4189–4197.

Hu, J., Zhang, H., Zhai, Y., 2006. Performance degradation studies on PBI/H_3PO_4 high temperature PEMFC and one-dimensional numerical analysis. Electrochim. Acta 52, 394–401.

Jalani, N.H., Ramani, M., Ohlsson, K., Buelte, S., Pacico, G., Pollard, R., et al., 2006. Performance analysis and impedance spectral signatures of high temperature PBI-phosphoric acid gel membrane fuel cells. J. Power Sources 160, 1096–1103.

Janßen, H., Supra, J., Luke, L., Lehnert, W., Stolten, D., 2013. Development of HT-PEFC stacks in the kW range. Int J Hydrogen Energy 38, 4705–4813.

Jensen J.O., Li Q., Terkelsen C., Rudbech H.C., Steenberg T., Rycke T. 2007. Development of HT-PEMFC components and stack for CHP unit. <http://goo.gl/n3l6Lx> (accessed May 2016).

Jiao, K., Li, X., 2010. A three-dimensional non-isothermal model of high temperature proton exchange membrane fuel cells with phosphoric acid doped polybenzimidazole membranes. Fuel Cells 10, 351–362.

Karan K., 2007. Structural modeling of PEMFC anode. In: Proceedings of 211th Meeting of the Electrochemical Society, 6–10 May, Chicago, IL.

Korsgaard, A.R., Refshauge, R., Nielsen, M.P., Bang, M., Kær, S.K., 2006a. Experimental characterization and modeling of commercial polybenzimidazole-based MEA performance. J. Power Sources 162, 239–245.

Korsgaard A.R., Refshauge R., Bang M., Kær S.K., 2006b. Modeling of CO influence in PBI electrolyte PEM fuel cells. In: The 4th International Conference on Fuel Cell Science, Engineering and Technology, 19–21 June, Irvine, CA.

Korsgaard, A.R., Nielsen, M.P., Kær, S.K., 2008a. Part one: a novel model of HTPEM-based micro-combined heat and power fuel cell system. Int. J. Hydrogen Energy 33, 1909–1920.

Korsgaard, A.R., Nielsen, M.P., Kær, S.K., 2008b. Part two: control of a novel HTPEM-based micro combined heat and power fuel cell system. Int. J. Hydrogen Energy 33, 1921–1931.

Kulikovsky, A.A., McIntyre, J., 2011. Heat flux from the catalyst layer of a fuel cell. Electrochim. Acta 56, 9172–9179.

Kulikovsky, A.A., Oetjen, H.F., Wannek, C., 2010. A simple and accurate method for high temperature PEM fuel cell characterisation. Fuel Cells 10, 363–368.

Kvesic, M., Reimer, U., Froning, D., Luke, L., Lehnert, W., Stolten, D., 2012a. 3D modeling of a 200 cm2 HT-PEFC short stack. Int. J. Hydrogen Energy 37, 2430–2439.

Kvesic, M., Reimer, U., Froning, D., Luke, L., Lehnert, W., Stolten, D., 2012b. 3D modeling of an HT-PEFC stack using reformate gas. Int. J. Hydrogen Energy 37, 12438–12450.

Lobato, J., Canizares, P., Rodrigo, M.A., Pinar, F.J., Mena, E., beda, D.U., 2010a. Three-dimensional model of a 50 cm² high temperature PEM fuel cell. Study of the flow channel geometry influence. Int. J. Hydrogen Energy 35, 5510–5520.

Lobato, J., Canizares, P., Rodrigo, M.A., Piuleac, C.G., Curteanu, S., Linares, J.J., 2010b. Direct and inverse neural networks modelling applied to study the influence of the gas diffusion layer properties on PBI-based PEM fuel cells. Int. J. Hydrogen Energy 35, 7889–7897.

Lobato, J., Canizares, P., Rodrigo, M.A., Pinar, F.J., Ubeda, D., 2011. Study of flow channel geometry using current distribution measurement in a high temperature polymer electrolyte membrane fuel cell. J. Power Sources 196, 4209–4217.

Mamlouk, M., Sousa, T., Scott, K., 2011. A high temperature polymer electrolyte membrane fuel cell model for reformate gas. Int. J. Electrochem. 2011, 1–18.

McConnell, V.P., 2009. High-temperature PEM fuel cells: hotter, simpler, cheaper. Fuel Cells Bulletin 2009, 12–16.

Oono, Y., Fukuda, T., Sounai, A., Hori, M., 2010. Influence of operating temperature on cell performance and endurance of high temperature proton exchange membrane fuel cells. J. Power Sources 195, 1007–1014.

Park, S.K., Choe, S.Y., 2008. Dynamic modelling and analysis of a 20-cell PEM fuel cell stack considering temperature and two-phase effects. J. Power Sources 179, 660–672.

Peng, J., Lee, S.J., 2006. Numerical simulation of proton exchange membrane fuel cells at high operating temperature. J. Power Sources 162, 1182–1191.

Peng, J., Shin, J.Y., Song, T.W., 2008. Transient response of high temperature PEM fuel cell. J. Power Sources 179, 220–231.

Rao, R.M., Bhattacharyya, D., Rengaswamy, R., Choudhury, S.R., 2007. A two-dimensional steady state model including the effect of liquid water for a PEM fuel cell cathode. J. Power Sources 173, 375–393.

Reddy, E.H., Jayanti, S., 2012. Thermal management strategies for a 1 kWe stack of a high temperature proton exchange membrane fuel cell. Appl. Therm. Eng. 48, 465–475.

Reddy, E.H., Monder, D.S., Jayanti, S., 2013. Parametric study of an external coolant system for a high temperature polymer electrolyte membrane fuel cell. Appl. Therm. Eng. 58, 155–164.

Samsun, R.C., Pasel, J., Janssen, H., Lehnert, W., Peters, R., Stolten, D., 2014. Design and test of a 5 kWe high-temperature polymer electrolyte fuel cell system operated with diesel and kerosene. Appl. Energy 114, 238–249.

Scholta, J., Zhang, W., Jörissenc, L., Lehnert, W., 2008. Conceptual design for an externally cooled HT-PEMFC stack. ECS Trans. 12, 113–118.

Scholta, J., Messerschmidt, M., Jörissen, L., Hartnig, Ch, 2009. Externally cooled high temperature polymer electrolyte membrane fuel cell stack. J. Power Sources 190, 83–85.

Scott, K., Mamlouk, M., 2009. A cell voltage equation for an intermediate temperature proton exchange membrane fuel cell. Int. J. Hydrogen Energy 34, 9195–9202.

Scott, K., Pilditch, S., Mamlou, M., 2007. Modelling and experimental validation of a high temperature polymer electrolyte fuel cell. J. Appl. Electrochem. 37, 1245–1259.

Shamardina, A., Chertovich, A., Kulikovsky, A.A., Khokhlov, A.R., 2010. A simple model of a high temperature PEM fuel cell. Int. J. Hydrogen Energy 35, 9954–9962.

Shamardina, A., Kulikovsky, A.A., Chertovich, A.V., Khokhlov, A.R., 2012. A model for high temperature PEM fuel cell: the role of transport in the cathode catalyst layer. Fuel Cells 12, 577–582.

Siegel, C., Bandlamudi, G., Heinzel, A., 2011. Systematic characterization of a PBI/H$_3$PO$_4$ sol-gel membrane-modeling and simulation. J. Power Sources 196, 2735–2749.

Song, T.W., Choi, K.H., Kim, J.R., Yi, J.S., 2011. Pumpless thermal management of water-cooled high-temperature proton exchange membrane fuel cells. J. Power Sources 196, 4671–4679.

Sousa, T., Mamlouk, M., Scott, K., 2010a. An isothermal model of a laboratory intermediate temperature fuel cell using PBI doped phosphoric acid membranes. Chem. Eng. Sci. 65, 2513–2530.

Sousa, T., Mamlouk, M., Scott, K., 2010b. A non-isothermal model of a laboratory intermediate temperature fuel cell using PBI doped phosphoric acid membranes. Fuel Cells 10, 993–1012.

Sousa, T., Mamlouk, M., Rangel, C.M., Scott, K.M., 2012. Three dimensional model of a high temperature PEMFC. Study of the flow field effect on performance. Fuel Cells 12, 566–576.

Song, G.H., Meng, H., 2013. Numerical modeling and simulation of PEM fuel cells: progress and perspective. Acta Mech. Sin. 29, 318–334.

Srinivasrao, M., Bhattacharyya, D., Rengaswamy, R., Narasimhan, S., 2010. Performance analysis of a PEM fuel cell cathode with multiple catalyst layers. Int. J. Hydrogen Energy 35, 6356–6365.

Sun, W., Brant, A.P., Karana, K., 2005. An improved two-dimensional agglomerate cathode model to study the influence of catalyst layer structural parameters. Electrochim. Acta 50, 3359–3374.

Tao, W.Q., Min, C.H., Liu, X.L., He, Y.L., Yin, B.H., Jiang, W., 2006. Parameter sensitivity examination and discussion of PEM fuel cell simulation model validation: Part I. Current status of modeling research and model development. J. Power Sources 160, 359–373.

Ubong, E.U., Shi, Z., Wang, X., 2009. Three-dimensional modeling and experimental study of a high temperature PBI-based PEM fuel cell. J. Electrochem. Soc 156, B1276–B1282.

Yu, S.H., Sohn, S., Nam, J.H., Kim, C.J., 2009. A numerical study to examine the performance of multi-pass serpentine flow fields for cooling plates in polymer electrolyte membrane fuel cells. J. Power Sources 194, 697–703.

Zhang, G., Kandlikar, S.G., 2012. A critical review of cooling techniques in proton exchange membrane fuel cell stacks. Int. J. Hydrogen Energy 37, 2412–2429.

Zhang, J., Tang, Y., Song, C., Zhang, J., 2007. Polybenzimidazole membrane based PEM fuel cell in the temperature range of 120–200°C. J. Power Sources 72, 163–171.

CHAPTER 5

Stationary HT-PEMFC-Based Systems—Combined Heat and Power Generation

5.1 INTRODUCTION

Combined heat and power (CHP) systems are being considered and used more frequently and are becoming more popular in modern society. The systems cover both the electrical and thermal load demands of a household and use fuel supplied from existing gas infrastructure. The electrical energy production surplus might be exported to the electrical grid and the surplus of heat might be used for hot water preparation (Fig. 5.1).

The introduction of cogeneration (generation of both electrical energy and heat) into the current energy sector offers many benefits to end users as well as to the environment. With the aid of the CHP system electrical energy and heat can be generated in a decentralized way, at the point of use, decreasing the losses related to electricity or heat distribution. Moreover, decentralized electricity production ensures an uninterrupted power supply which is not affected by the faults of the distribution grid. Electrical energy and heat are produced in a cogenerated manner from a single source of fuel and used at the point of generation, thus improving the total efficiency of such a process. By the application of CHP technology and generation of power in a decentralized way, the efficiency of the process can be significantly increased. The typical efficiency of the energy conversion process using fossil fuel is about 35–40%, and can be increased to 80–90% by using fuel cell–based CHP (FC-CHP) systems (Fig. 5.2).

In this way the amount of fuel required to generate a given amount of energy decreases, thus reducing CO_2 emissions as well as significantly cutting energy production costs. The outputs of recent studies performed by Berger (2015) have shown that FC-CHP systems might save up to 27% of the typical European home energy consumption and CO_2 emissions caused by the energy generation process might be

Recent Advances in High-Temperature PEM Fuel Cells. DOI: http://dx.doi.org/10.1016/B978-0-12-809989-6.00005-0
© 2016 Elsevier Ltd. All rights reserved.

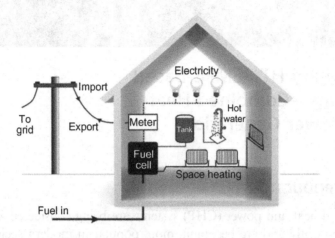

Figure 5.1 Residential CHP system based on a fuel cell. Reproduced from Hawkes, A., Staffell, I., Brett, D., Brandon, N., 2009. Fuel cells for micro-combined heat and power generation. Energy Environ. Sci. 2, 729–744 (Hawkes et al., 2009).

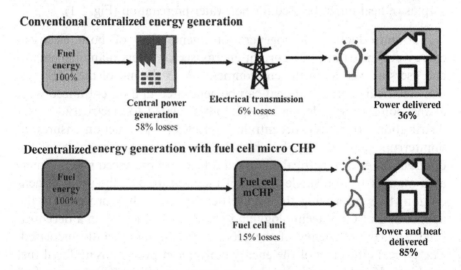

Figure 5.2 Comparison of efficiency of energy generation process using conventional centralized method and decentralized FC-CHP system by Elmer et al. (2015).

decreased to about 30%. Comparison studies by Slowe and Elmer et al. (2015) show the advantage of FC-CHP systems over other CHP technologies in the power range up to 1 kW. It was shown that solid oxide fuel cell (SOFC)- and polymer electrolyte membrane fuel cell (PEMFC)-based systems have the best emission performance (Fig. 5.3).

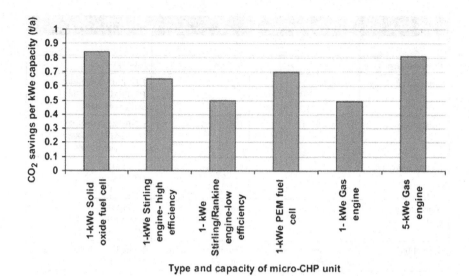

Figure 5.3 Annual CHP CO₂ savings compared to grid electricity and boiler alternatives; by Slowe, available in Elmer et al. (2015).

CHP systems that use internal combustion engines (ICE) and Stirling engines (SE) are currently available, but FC-CHP systems are being developed as the fuel cell technology becomes more mature and commercially available (Table 5.1).

As reported by Dwyer (2012), the number of sales of CHP systems increases alongside progress in technology development. The number of FC-CHP systems is increasing steadily and the new state-of-the-art technology is taking over units based upon the conventional technologies such as ICE or SE (Fig. 5.4).

According to the data collected and compiled by Dodds et al. (2015) and based upon prepared forecasts, the number of installed FC-CHP systems will increase on a yearly basis and 100% year-on-year growth is predicted. The trend shows that the number of installed systems will reach 1,000,000 in Japan and 100,000 in Europe by 2020 (Fig. 5.5).

The world leader in the commercialization of FC-CHP units is Japan. Recently Horisaka (2016) reported cumulative sales that exceeded 140,000 units, a value which is in close agreement with that forecast by Dodds et al. (2015), as shown in Fig. 5.6. A stable increase in unit sales has been observed after 2 years from the implementation

Table 5.1 Selected CHP Technologies

	ICE	SE	FC
Electrical capacity	1–5 kW	1–5 kW	0.3–5 kW
Electrical efficiency	20–30%		LT-PEMFC—30–40% HT-PEMFC—~30% SOFC—40–60%
Total efficiency	Up to 90%	Up to 95%	LT-PEMFC—80–90% HT-PEMFC—~85% SOFC—up to 90%
Heat to power ratio	3	6	LT-PEMFC—2 HT-PEMFC—2 SOFC—1
Ability to vary the load	No	No	LT-PEMFC—yes HT-PEMFC—yes SOFC—no
Type of fuel	Natural gas, biogas, liquid fuels	Natural gas, biogas, butane	Natural gas, city gas, LPG, hydrogen
Noise	Loud	Fair	Quiet
Maturity	High	Fair	Fair
Producers			See Table 5.3

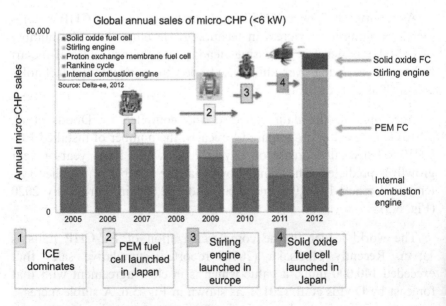

Figure 5.4 Global annual sales of micro-CHP (≤6 kW).

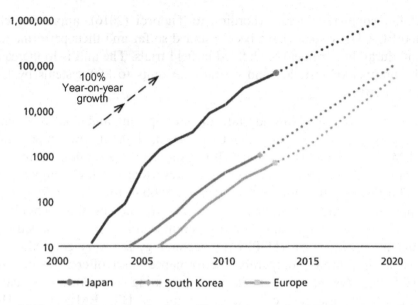

*Figure 5.5 Historic (*solid lines*) and predicted (*dotted lines*) cumulative number of FC-CHP systems installed in three major regions, as compiled by Dodds et al. (2015).*

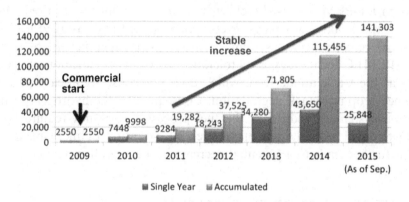

Figure 5.6 Cumulative and annual FC-CHP unit sales in Japan.

of the plan for commercialization that was introduced in 2009. In 2012 the Japanese government announced very ambitious targets for cumulative installations of stationary FC-CHP systems in the residential sector, with plans to reach over 1.4 million units installed by the end of 2020 and 5.3 million units by 2030.

The market for the stationary application of FC-CHP systems in the residential sector is currently being established by a few European

Union countries. Here, according to Tudoroi (2016), approximately 400 FC-CHP systems have been installed so far and their performance and durability are being validated in field trials. The aim is to increase the number of installed and operational units to 860 systems by the end of 2016.

Owing to the advanced state of development and relatively high technology readiness level (TRL) of about 4–6 of low-temperature PEMFCs (LT-PEMFCs) and SOFCs, those two technologies are currently used in FC-CHP systems. Nevertheless, high-temperature (HT-PEMFC) technology has great potential for FC-CHP system applications and advanced developments are moving this technology from the research phase toward commercialization. As stated by Ellamla et al. (2015), HT-PEMFC technology is a very good alternative as it exhibits comparable performances and offers system cost reduction. Owing to the elevated operating temperature (usually >120°C) of the fuel cell stack, the use of HT-PEMFCs in CHP systems simplifies the system layout and thus decreases system complexity. Some of the system components that are crucial for operation in LT-PEMFC-based fuel cell systems are no longer required. One may expect that fewer parts or components used and installed lead to a more reliable overall system operation. The other benefit of using HT-PEMFC technology is its tolerance for CO concentration of up to 3% at the stack anode feed, which does not cause significant performance drop nor stack degradation. Consequently, the hydrogen purification system embedded in the fuel processor does not require final CO removal with the aid of preferential oxidation or methanization. This further leads to a potential decrease in the overall system cost.

5.2 HT-PEMFC-BASED FC-CHP SYSTEM

The principle of operation of HT-PEMFC-based FC-CHP systems is based upon the same concept and is very similar to LT-PEMFC- or SOFC-based systems, as shown in Fig. 5.7. For HT-PEMFC-based FC-CHP systems, natural gas is used as primary fuel and converted via a steam methane reformer (SMR) into hydrogen-rich reformate gas composed, depending on the complexity of the gas purification system, of CH_4 <2%, CO <1%, CO_2 ~20%, and H_2 ~78%. The reformate gas is fed into the fuel cell stack where the chemical energy of the hydrogen is converted into electricity and heat. DC electricity is further

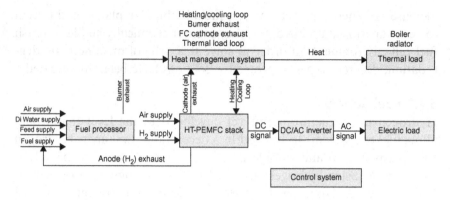

Figure 5.7 Schematic presentation of HT-PEMFC-based FC-CHP system.

processed by a series of DC/DC converters and DC/AC inverters to supply useful power to the end user. The anode off-gas is directed into the fuel processor burner to provide the fuel required to heat up the reformer and run reforming reactions. Waste heat is captured from the fuel processor, stack cooling loop and cathode exhaust and stored in heat storage or used to cover the heat demands of the end user.

5.2.1 Stack

Because of the elevated operating temperatures, the typical state-of-the-art fuel cell materials used for LT-PEMFC stack construction cannot be applied in HT-PEMFC technology. This is why the stack components such as membrane electrode assemblies (MEAs), including membranes and gas diffusion electrodes (GDE), gaskets, and bipolar plates, as well as system components, need to be developed to allow operations at temperatures up to $\sim 200°C$. The widely used LT-PEMFC Nafion® membrane material has to be substituted with a material in which the ionic conductivity does not depend upon the water content as at high temperature the electrolyte would dry out and lose its important properties. To cater for this requirement, acid-based polymer electrolytes, e.g., phosphoric acid–doped polybenzimidazole (PBI)-type materials are commonly used. For PBI-based electrolytes, proton conductivity depends upon the phosphoric acid content within the material, which most of the time causes problems during the operation, as the acid tends to leach out. It is also for that reason that bipolar plates used for HT-PEMFC are manufactured from high-temperature versions of graphite–resin composite plates owing to their chemical stability and high electrical conductivity. To reduce costs and increase volumetric power density, metallic bipolar plates might seem to be a superior alternative, but they are not chemically stable in harsh fuel cell environment. In order to solve this issue of metallic plate degradation, different types of protective coatings have been investigated.

5.2.2 Fuel Supply

Hydrogen infrastructure is not yet developed to the level that would allow fuel cell systems to be used in a decentralized manner. Hydrogen gas transportation and storage presents another problem to be solved in enabling faster hydrogen and fuel cell technology commercialization. In the meantime on-site hydrogen production strategies for fuel cell supply are being used. In this case the hydrogen production, at the point of installation of the system, is realised using a fuel processor and natural gas, methanol or diesel as a primary fuel. The steam methane reforming (SMR) method is often used as the process for hydrogen production in which the reformate gas is produced with the aid of autothermal reformers or partial oxidation reactors.

As shown in Fig. 5.8, published by Steele (1999), fuel processor complexity increases with decreasing fuel cell operating temperatures. HT-PEMFC technology, with its nominal operating temperature of about 160–180°C and its close similarity to phosphoric acid fuel cell (PAFC) technology, does not require complex hydrogen purification and can tolerate, without significant influence on performance, up to

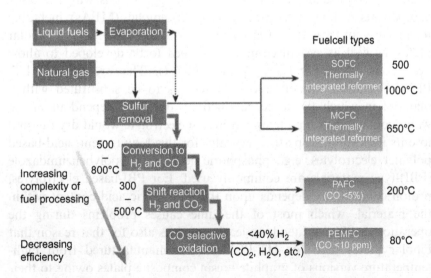

Figure 5.8 Reforming of methane to hydrogen fuel for application with different types of fuel cells as presented by Steele (1999).

3% of CO in the anode gas feed. As the complexity of the fuel processing system increases, the efficiency of the overall process decreases, as more energy is lost owing to the additional reactions occurring to clean the reformate gas.

5.2.3 Thermal Management and Heat Recuperation

Because of the high operating temperatures of the HT-PEMFC stack, various approaches have to be taken into account in order to ensure an efficient stack cooling system and effective heat dissipation or recuperation, which accompanies the generation of electrical energy. Based upon the type of media used to cool the stack and maintain its nominal operating temperature, air or liquid cooling systems are employed. Both methods have advantages and disadvantages. Air-cooled systems are simpler and easy to realize, but they require high levels of air flow and more importantly, the efficiency of heat dissipation is quite low. The flow of cooling air can be forced through open or closed stack cathode or separate cooling plates with cooling channels. Liquid-cooled systems are more suitable for HT-PEMFC applications but they are not free of problems. Because the operating temperatures are much higher than the boiling point of water, the use of this liquid is limited. Liquid-cooled systems operating with water at a temperature of 160°C have to be pressurized to about 10 bar to prevent water boiling. A solution to this is the application of different liquids/solutions, e.g., thermal oil with a higher boiling point. Thermal oils possess very low electrical conductivity and do not cause corrosion of the bipolar plates, but they react when in contact with rubber-based sealing materials, thus leading to MEA contamination. High oil viscosity, especially at low temperatures, has also to be addressed to avoid large pressure drops in the loop during stack preheating from room temperature. To avoid these problems, Scholta et al. (2009) proposed externally oil-cooled stack construction. In this type of cooling system, the oil is circulated in an external loop and has no contact with the stack components.

Besides the fuel cell stack cooling system, heat is generated in other places of the FC-CHP system and can be recuperated with the aid of heat exchangers to increase system efficiency. These places include reformer output, where gas is available at 200–250°C; burner exhaust, from where the heat can be recuperated from the flue gas that leaves the combustion chamber at about 250°C, as well as from the cathode off-gas that exits the HT-PEMFC stack at the stack operating temperature of about 160°C. The high-quality heat recuperated from these places in the form of hot water or/and low pressure steam can be used within the system to preheat fuel cell reactants or externally for heating purposes and hot-water preparation.

5.2.4 Power Conditioning

The HT-PEMFC stack polarizations have nonlinear characteristics and the DC stack or cell voltage changes in a wide range, as the open circuit voltage and operating cell voltage may vary significantly. Thus the use of some type of power conditioning device is required. Usually, in the first step, unregulated stack cell voltage is controlled to a constant value of $+12$, $+24$, or $+48$ V with the aid of a DC/DC converter. Higher converter output voltages are preferable, otherwise high-current devices have to be used for the same power. High-current electronic components generate more heat that has to be dissipated, which may significantly affect overall system efficiency. If the fuel cell cell voltage is lower than required by the system, a boost (step-up) DC converter is used and in the opposite situation a buck (step-down) DC converter needs to be employed. It is also possible to omit DC/DC converters in the construction of fuel cell systems. In this case the fuel cell operating cell voltage has to be carefully aligned with the voltage required by the system. Finally, in the case of stationary systems, in order to supply end users with power that could be used to supply electrical appliances a DC/AC inverter has to be integrated within the system. To cover peak power demands and to supply auxiliaries during system start-up and shut-down, small energy storage in the form of battery packs or a supercapacitors bank is usually a part of the system.

5.2.5 System Control and Operation

In the case of HT-PEMFC systems, the number of auxiliaries that need to be controlled is smaller, but nevertheless, a control system that supervises the operation of all auxiliaries is necessary. Typically, Programmable Logic Controllers (PLCs) with I/O modules or microcontroller chips with programed operation strategy are used to manage system operation.

The CHP system can operate in either an electric load following mode or a thermal load following mode. It is quite intuitive, but to explain in more detail, in an electric load following mode the system responds to the electric load demand and the thermal energy is recuperated as a byproduct that can be used, stored or dumped into the atmosphere. In the second case thermal load demand drives the system operation and electrical energy is generated as a byproduct that can be used onsite or can be returned into the grid if the system is grid-connected. In the case of grid-independent areas (e.g., remote areas such as islands) CHP system installation requires a mix thermal/electric load demand operation strategy.

5.2.6 System Costs

Decreasing the cost of FC-CHP systems is a great challenge. For PEMFC-based systems the cost reduction might be achieved by decreasing fuel processor costs, which currently constitutes 80% of the total Balance of Plant (BoP) components' cost. Based on the status of the technology in 2008, the targeted factory cost of stationary FC-CHP systems operating on natural gas, established by the US Department of Energy (DOE) and published in the report by Maru et al. (2010), was about $550 per kW in 2015, taking into account a yearly production of 50,000 units with 5 kW power, with a target cost of under $450 per kW by 2020. In the same report the stakeholders acknowledged that the established target is very challenging but could be achievable for high-volume production (50,000 units). Nevertheless at lower volumes, closer to 5,000 units per year, the costs would be doubled. The technology targets were revisited based on the status of the technology in 2011 when the cost of the FC-CHP system was in the range of $2,300–4,000 per kW. The targets have seen many reiterations as they have been following the status of technology development and pricing. The latest update (Table 5.2) published by US DOE (2016) includes an increase of targeted factory costs to $1,500 per kW by 2020, established for high-volume production of 5 kW systems.

In the current state of deployment, the capital cost of FC-CHP systems is currently substantially higher than the US DOE cost targets. However, early commercialization of the technology and the implementation of demonstration programs have seen an increase in the number of manufactured and installed units, leading to a decrease in price to the level at which the new technology is now becoming competitive with currently used solutions. According to Staffell and Green (2009), the mass production and the doubling of the number of manufactured units will enable reduction of the price in the range of 9.1–21.4%. Recently Horisaka (2016) confirmed the system price reduction pointed out by Staffell et al. as shown in Fig. 5.9. In 2016 the recommended retail price

Table 5.2 Technical Targets: Residential Combined Heat and Power and Distributed Generation Fuel Cell Systems Operating on Natural Gas[a] Compiled Based on Maru et al. (2010), US DOE (2012), and US DOE (2016)

Characteristic	2008 Status	2012 Target	2011 Status	2015 Target	2015 Status	2020 Targets
Electrical efficiency at rated power[b]	34%	40%	34–40%	42.5%	34–40%	>45%[c]
CHP energy efficiency[d]	80%	85%	80–90%	87.5%	80–90%	90%
Equipment cost[e], 2-kW$_{avg}$[f] system	NA	NA	NA	US $1,200 per kW$_{avg}$	NA	US $1,000 per kW$_{avg}$
Equipment cost[e], 5-kW$_{avg}$ system	US$750 per kW[g]	US$650 per kW[g]	US $2,300–4000 per kW[h]	US $1,700 per kW$_{avg}$[h]	US $2,300–2,800 per kW[i]	US $1,500 per kW$_{avg}$[i]
Equipment cost[e], 10-kW$_{avg}$ system	NA	NA	NA	US $1,900 per kW$_{avg}$	NA	US $1,700 per kW$_{avg}$
Transient response (10–90% rated power)	5 min	4 min	5 min	3 min	5 min	2 min
Start-up time from 20°C ambient temperature	60 min	45 min	<30 min	30 min	10 min	20 min
Degradation with cycling[j]	<2% per 1,000 h	0.7% per 1,000 h	<2% per 1,000 h	0.5% per 1,000 h	<2% per 1,000 h	0.3% per 1,000 h
Operating lifetime[k]	6,000 h	30,000 h	12,000 h	40,000 h	12,000–70,000 h	60,000 h
System availability[l]	97%	97.5%	97%	98%	97%	99%

[a]Pipeline natural gas delivered at typical residential distribution line pressures.
[b]Regulated AC net/LHV of fuel.
[c]Higher electrical efficiencies (e.g., 60% using SOFC) are preferred for non-CHP applications.
[d]Ratio of regulated AC net output energy plus recovered thermal energy to the LHV of the input fuel. For inclusion in CHP energy efficiency calculation, heat must be available at a temperature sufficiently high to be useful in space- and water-heating applications. Provision of heat at 80°C or higher is recommended.
[e]Complete system, including all necessary components to convert natural gas to electricity suitable for grid connection, and heat exchangers and other equipment for heat rejection to conventional water heater, and/or hydronic or forced air heating system. Includes all applicable tax and markup. Based on projection to high-volume production (50,000 units year^{-1}).
[f]kW$_{avg}$ is the average output (AC) electric power delivered over the life of a system while the unit is running.
[g]Cost includes materials and labor costs to produce the stack, plus any balance of plant necessary for stack operation. Cost defined at 50,000 unit year^{-1} production (250 MW in 5-kW modules).
[h]Strategic Analysis, Inc. preliminary 2011 cost assessment of stationary PEM system, range represents manufacturing volumes of 100–50,000 units year^{-1}.
[i]Battelle preliminary 2015 cost assessment of stationary CHP systems, range represents different technologies (SOFC vs PEMFC) at manufacturing volumes of 50,000 units year^{-1}.
[j]Durability testing should include effects of transient operation, start-up, and shutdown.
[k]Time until >20% net power degradation.
[l]Percentage of time the system is available for operation under realistic operating conditions and load profile. Unavailable time includes time for scheduled maintenance.

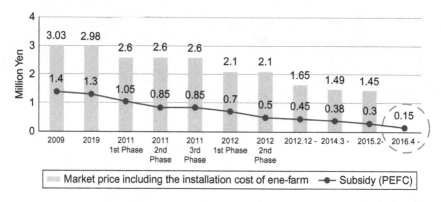

Figure 5.9 FC-CHP system cost and subsidy amount in Japan by Horisaka (2016).

of a FC-CHP system offered by *Panasonic* was about US$3,000 (~¥300,000) lower than that of a 2013 model which had the same basic functionalities. It was found that the trend of price reduction is observed with increasing numbers of installed FC-CHP systems and currently the price of the system on the Japanese market is about US$15,000 (¥1.45 M), i.e., a cost of US$22,000 per kW.

If the number of installed FC-CHP systems increases on a yearly basis, one may expect an attractive price reduction (this forecast is based on historical prices, as presented by Dodds et al., 2015). Thus one may assume that the price of a complete FC-CHP system will be in the range of US$6,500−13,000 (£4,500−9,000) (Fig. 5.10).

One of the main technical drivers for the price reduction might be reducing system complexity and eliminating major system components such as fuel processing and purification stages. This may be achieved by using HT-PEMFC technology in FC-CHP systems.

5.3 FC-CHP SYSTEMS, DEMONSTRATION PROGRAMS, AND HT-PEMFC-BASED SYSTEMS

It is evident that the technology of FC-CHP systems has reached stage of development that enables the commercialization phase to commence. Currently a few manufacturers offer commercial FC-CHP systems and there are still ongoing R&D projects being undertaken by research communities that focus on the development of new materials/components and prototypes/products. Moreover, the output power of most systems does not exceed 1 kW for residential applications (as this

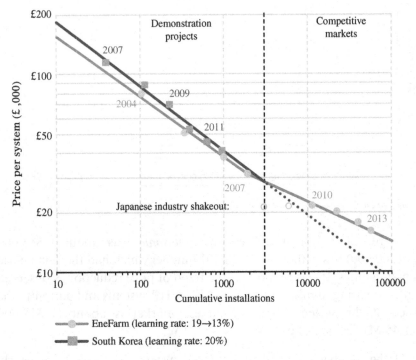

Figure 5.10 Historic prices of Japanese Ene-Farm and South Korean residential PEMFC-based systems compiled by Dodds et al. (2015).

power fulfills the demands of a single household). The companies that are working in the field of FC-CHP systems are located in various places around the world, but once again Japan and European countries, e.g., Germany, take the lead. In general two fuel cell technologies, PEMFC and SOFC, are used but HT-PEMFC-based systems are under development (Table 5.3).

5.3.1 Ene-farm Program

Japan has the largest fuel cell commercialization program and is the world leader in terms of the number of installed FC-CHP systems. In the framework of the *Ene-farm* program, over 140,000 units have been installed. New models of FC-CHP systems for apartment buildings and homes were introduced to the market in 2015 (Fig. 5.11). The new systems are smaller, more efficient, cost-effective and more easily installed than previously available units. Currently the main participants of the program are *Panasonic* and *Toshiba*, with LT-PEMFC-based units on the market, and *Aisin Seiki* offering SOFC-based FC-CHP systems.

Table 5.3 Updated Table of FC-CHP Systems Available on the Market or in Development Stages Compiled by Elmer et al. (2015)

Manufacturer (Country of Origin[a])	FC Technology	Electrical Capacity (kWe)	Electrical Efficiency (%)	Thermal Output (kWth)	Auxiliary Heater Included	Cost	Commercial Availability	Partners/ Projects	Comments
Baxi (UK)	PEMFC	1	32	1.7	20 kWth	—	2015	Ballard/ Callux	Requires external heater
Toshiba (JPN)	PEMFC	0.7	35	1	—	US $20,000	Japan 2009 Europe 2015	Ene-Farm	80,000 operational hours expected
Viessmann (GER)	PEMFC	0.75	37	1.3	19 kWth	US $40,000 (€35,000)	Germany 2014 Europe 2015	Panasonic	Uses Japanese stack
Dantherm Power (DEN)	PEMFC	1.7 2.5 5	—	—	—		Danish field trials	Ballard	Only short duration tests so far
Panasonic (JPN)	PEMFC	0.7	40	0.9	—	US $28,000 (€25,000)	Japan 2011 Europe 2014	—	European R&D started in 2012
JX Eneos (JPN)	PEMFC	0.7	40	—	—	—	Japan 2011	Ene-Farm	Pursuing SOFC technology
Vaillant (GER)	PEMFC	5	—	25–50	—	—	—	Plug Power	Aimed at multifamily houses
Elcore (GER)	HT-PEMFC	0.3	33	0.6	—	US $9,000 (€10,000)	Ene.Field 2013	Ene.Field	Low electrical/heat output means the FC runs continuously
HySA Systems (RSA)	HT-PEMFC	1	30	—	—	—	HySA 2018	TFD/HySA	Laboratory testing ongoing, field testing in 2017
Plug Power (USA)	HT-PEMFC	0.3–3	30	1.65	—	—	—	—	Operates on natural gas
CFCL (AUS)	SOFC	1.5	60	0.6–1	No	US $29,000 (£20,000)	Yes	E.On	Highest electrical efficiency on the market

(Continued)

Table 5.3 (Continued)

Manufacturer (Country of Origin[a])	FC Technology	Electrical Capacity (kWe)	Electrical Efficiency (%)	Thermal Output (kWth)	Auxiliary Heater Included	Cost	Commercial Availability	Partners/Projects	Comments
Hexis (SUI)	SOFC	1	30–35	1.8	20 kW	—	Callux 2012	Vissmann/Callux	Electrical efficiency similar to PEMFC
Ceres Power (UK)	SOFC	1	—	—	—	—	2016	British Gas/KD Navien	External reformer
Vaillant (GER)	SOFC	1	30	1.7	—	—	2013	Staxera/Callux	Focus on reliability
Kyocera (JPN)	SOFC	0.7	46.5	0.65	—	—	Japan 2012	Osaka Gas	Uses flat tubular cells
Aisin Seiki (JPN)	SOFC	0.7	46.5	—	—	US $30,500 (£21,000)	Japan 2012	Osaka Gas/Bosch	—
JX Eneos (JPN)	SOFC	0.7	45	—	40 kWth	US $31,000	Japan 2012	Kyocera	Robust unit
Topsoe (DEN)	SOFC	1	—	—	—	—	—	Wartsila/Dantherm	Robust cells
Acumentrics (USA)	SOFC	0.25–1.5	<35	—	—	—	2013	—	Able to respond to thermal cycling
SOFC power (SUI)	SOFC	0.5/1	30–32	—	—	—	—	—	Low electrical efficiency for SOFC
Acumentrics (USA)	SOFC	1/2.5 peak	30	—	Up to 24 kW	—	Not in general availability	—	Operates on natural gas

[a] AUS—Australia, DEN—Denmark, GER—Germany, JPN—Japan, RSA—Republic of South Africa, SUI—Switzerland, UK—United Kingdom, USA—United States of America.

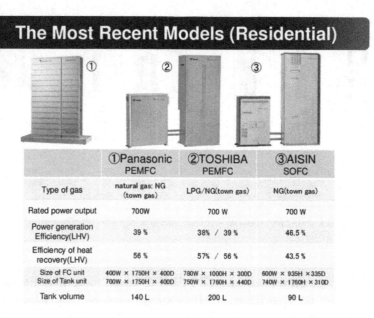

	①Panasonic PEMFC	②TOSHIBA PEMFC	③AISIN SOFC
Type of gas	natural gas: NG (town gas)	LPG/NG(town gas)	NG(town gas)
Rated power output	700W	700 W	700 W
Power generation Efficiency(LHV)	39 %	38% / 39 %	46.5 %
Efficiency of heat recovery(LHV)	56 %	57% / 56 %	43.5 %
Size of FC unit Size of Tank unit	400W × 1750H × 400D 700W × 1750H × 400D	780W × 1000H × 300D 750W × 1760H × 440D	600W × 935H × 335D 740W × 1760H × 310D
Tank volume	140 L	200 L	90 L

Figure 5.11 Specifications of FC-CHP systems available on Japanese market.

5.3.2 ene.field Program

The *ene.field* is a small-scale demonstration program based in Europe, aiming at FC-CHP systems commercialization. The main objective of the program is to install 1000 FC-CHP units in 11 European countries to demonstrate the environmental and economic potential/impact of FC-CHP technology in the residential sector. In this program European FC-CHP producers (*Baxi Innotech, Bosch, Dantherm Power, Elcore, Hexis, RBZ, Solid Power, Vaillant, and Viessmann*) offer different FC-CHP systems, which include LT-PEMFC, HT-PEMFC, and SOFC (and thus have different electrical and thermal outputs) as shown in Fig. 5.12.

5.3.3 HT-PEMFC-Based Systems

HT-PEMFC technology gives better flexibility with regard to the range of powers that can be supplied by the system, which is directly related to the size of the fuel cell stack used. Thus systems and units ranging from hundreds of watts to kilowatts of electric power can be manufactured and are currently available on the market. Another great advantage of HT-PEMFC-based systems is the possibility of supplying them with a wide range of fuels, e.g., natural gas, city gas, biogas, or even methanol. This is very attractive feature, because there is no hydrogen distribution infrastructure available.

Dachs InnoGen	Cerapower FC10 Logapower FC10	PEMmCHP G5	Elcore 2400	Galileo 1000 N	Inhouse 5000+	ENGEN 2500	BLUEGEN	Vaillant G5+	Vitovalor
LT PEM	SOFC	LT PEM	HT PEM	SOFC	LT PEM	SOFC	SOFC	SOFC	PEM
700W	700W	2kW	300W	1kW	5kW	2.5kW	2kW	1kW	700W
Natural Gas	Natural Gas, Gas	Natural Gas + Biogas	Natural Gas	Natural gas+ Biogas	Natural gas + Biogas + H2	Natural Gas	Natural Gas	Natural Gas	Natural Gas
Floor	Floor	Floor	Wall	Floor	Floor	Floor	Floor	Wall	Floor
SenerTec	Bosch Thermotechnik	Dantherm Power	Elcore	Hexis	RBZ	Solid power	Solid power	Vaillant	Viessmann

Figure 5.12 Models of FC-CHP systems developed and installed in the framework of the ene.field program.

The first example of an HT-PEMFC-based system is Elcore 2400, which is a 300 W μ(micro)-CHP system developed by *Elcore GmbH*, used for residential applications (Fig. 5.13). The system can provide up to 0.3 kW electrical power produced with 32% efficiency and 0.7 kW thermal energy generated at 72% efficiency for a household. The natural gas-fueled system can cover 50−70% of a household's annual electrical energy demands and 100% of its hot water requirements. The system is equipped with a reformer that converts natural gas into hydrogen-rich reformate; the water for steam reforming comes from the condensation of exhaust gases, increasing the system's total efficiency. The installation and operation of the system saves around 2 tons of CO_2 emission annually, which is the same amount as for a household powered by renewable energy sources.

The fuel for HT-PEMFC stacks can be produced onsite using methanol. The fuel composition is typically 60% methanol and 40% deionized water (60% CH_3OH:40% H_2O). This concept is used in the series of *H3 modules* developed by Serenergy that produce powers ranging from 0.35 kW to 5 kW with efficiency of up to 45% (Fig. 5.14). The use of methanol as a fuel yields high power density and high fuel energy density. The *H3 modules* can be used for automotive or backup power applications for charging a battery pack, externally or integrated within the system.

The development of FC-CHP systems for residential applications is one of the main objectives of the *Hydrogen South Africa (HySA)*

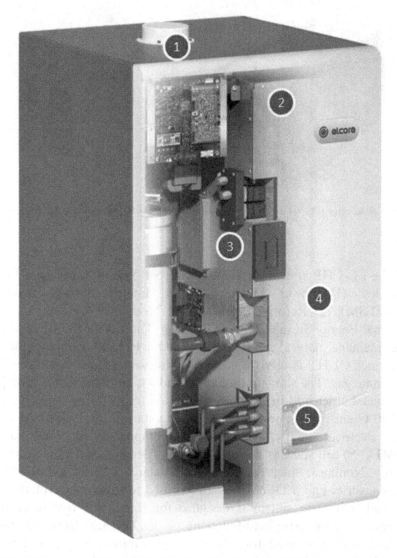

Figure 5.13 Construction of Elcore 2400 μ-CHP system: 1—exhaust, 2—HT-PEMFC stack, 3—exhaust gas water condensation, 4—integrated reformer for gas conversion, 5—service connections.

program. One of the aims of the R&D works undertaken at *HySA Systems Integration & Technology Validation Competence Centre (HySA Systems)* is to develop internationally competitive and marketable FC-CHP systems and critical FC-CHP system components. Recently a 1-kW HT-PEMFC FC-CHP system was developed and installed at *HySA Systems* laboratory (Fig. 5.15).

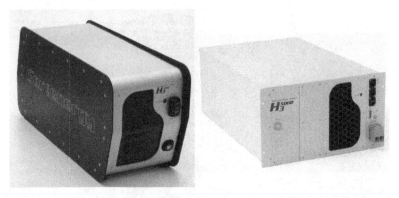

Figure 5.14 Serenergy H3 350 *methanol fuel cell generator (left) and* H3 5000 *methanol fuel cell generator (right).*

The FC-CHP system developed at *HySA Systems* consists of three modules: (1) Power Management & Energy Storage Module (PM&ESM), (2) Reformer & Fuel Cell Module (R&FCM), and (3) Thermal Energy Storage Module (TESM). The system can be supplied with methane, natural gas or city gas and a FLOX®-Compact Steam Reformer C1-HT is used to convert hydrocarbons into hydrogen-rich reformate gas. The start-up time is around 1 h and during this time the reformer burner and fuel cell are preheated to operating temperatures of 840°C and 140°C, respectively. During startup the PM&ESM supplies electrical energy to the system components. A 1-kW$_e$ HT-PEMFC 50-cell stack is integrated in the system to generate electrical energy. Nonlinear stack cell voltage is regulated with the aid of a DC/DC converter and a DC/AC inverter is used to supply AC voltage for the end user. The system is designed for standalone operation and operates in an electric load-following mode. The heat from the stack cooling system is used for hot water preparation and stored in a TESM. Two additional heat exchangers are installed to recuperate waste heat from the burner and fuel cell stack cathode outlet to increase overall system efficiency. *Siemens* PLC and *Weintek* Human Machine Interface (HMI) touch panel with data logging functionality are used for system control, communication, and data storage. The system undergoes characterization and testing in laboratory conditions.

The other example of HT-PEMFC technology used in stationary application is the FC-CHP system developed by *PlugPower*. The

Figure 5.15 FC-CHP system developed and installed at HySA Systems *laboratory.*

GenSys Blue, a natural gas-supplied 5 kW unit, was demonstrated in the framework of the project funded by the US DOE. The design assumptions offer 20−40% reduction in home energy costs and 25−35% reduction of carbon emissions. The aim of the project was to investigate internally (in the lab) and externally (on-site) the durability and reliability of *GenSys Blue* FC-CHP systems. According to the report by Tyler (2014) a fleet of 6 units was installed and tested in real-world residential end user locations in California (USA). During the course of the project, the installed systems operated for about 31,000 h and produced in total 52,905 kWh of electrical energy and generated a total of 632,998 kWh of heat at a total efficiency of 89%. The cost of US$0.06 per kWh was based on the natural gas price in California from December 2012. The payback time for the end user was predicted to be about 5–8 years.

5.4 SUMMARY

The application of fuel cell technology in stationary FC-CHP systems gives the possibility for decentralized energy production and better fuel utilization, owing to the systems' high levels of efficiency. The high capital costs of FC-CHP systems could be reduced by increasing the number of installed units and simplification of the system construction. HT-PEMFC-based FC-CHP systems, owing to their high operating temperatures, have a simpler layout of BoP devices, and moreover they can operate on reformate gas with relatively high CO content. Thus the gas purification subsystem of the reformer is much simpler and more cost-effective when the hydrogen is produced onsite from hydrocarbon fuels. Demonstration programs aimed at the installation and field testing of FC-CHP systems have been implemented and are running in Japan and Germany. Unfortunately, few companies worldwide are working on the development of HT-PEMFC technology for FC-CHP systems. The present results of R&D are promising; but nevertheless, more vigorous efforts are needed for the successful commercialization of HT-PEMFC-based FC-CHP systems.

REFERENCES

Berger, R., 2015. Advancing Europe's energy systems: stationary fuel cells in distributed generation. FCH JU – Fuel Cell Distributed Generation Commercialisation Study. Available from: <https://www.rolandberger.com/media/pdf/Roland_Berger_Fuel_Cells_Study_20150330.pdf> (last accessed March 2016).

Dodds, P.E., Staffell, I., Hawkes, A.D., Li, F., Grunewald, P., McDowall, W., et al., 2015. Hydrogen and fuel cell technologies for heating: a review. Int. J. Hydrog. Energy 40, 2065–2083.

Dwyer, S., 2012. Presentation delta energy & environment, cogeneration days 2012, European small scale cogen (sub-100kWe): market status & prospects. Available from: <www.deltaee.com/images/downloads/pdfs/Deltaee_mCHP_market_status_and_potential_Cogen_Czech_161012.pdf> (last accessed March 2016).

Ellamla, H.R., Staffell, I., Bujlo, P., Pollet, B.G., Pasupathi, S., 2015. Current status of fuel cell based combined heat and power systems for residential sector. J. Power Sources 293, 312–328.

Elmer, T., Worall, M., Wu, S., Riffat, S.B., 2015. Fuel cell technology for domestic built environment applications: state of-the-art review. Renew. Sust. Energy Rev 42, 913–931.

Hawkes, A., Staffell, I., Brett, D., Brandon, N., 2009. Fuel cells for micro-combined heat and power generation. Energy Environ. Sci. 2, 729–744.

Horisaka, K., 2016. The update on the status of residential fuel cell in Japan. In: Update Ene-Farm & Ene.Field Market Introduction Programs Webinar, 18 February 2016.

Maru, H., Singhal, S.C., Stone, C., Wheeler, D., 2010. 1–10 kW Stationary combined heat and power systems status and technical potential, NREL. Available from: <http://www.nrel.gov/docs/fy11osti/48265.pdf> (accessed 06.06.16).

Scholta, J., Messerschmidt, M., Jorissen, L., Hartnig, Ch, 2009. Externally cooled high temperature polymer electrolyte membrane fuel cell stack. J. Power Sources 190, 83–85.

Serenergy methanol fuel cell systems. <http://serenergy.com/applications/systems/> (accessed 08.04.16).

Staffell, I., Green, R.J., 2009. Estimating future prices for stationary fuel cells with empirically derived experience curves. Int. J. Hydrogen Energy 34, 5617–5628.

Steele, B.C.H., 1999. Fuel-cell technology: running on natural gas. Nature 400, 619–621.

The most effective fuel cells for electricity and heat – Elcore 2400. Available from: <http://www.elcore.com/produkte/elcore-2400/> (accessed 08.04.16).

Tudoroi, A., 2016. Ene.field: European-wide deployment of residential fuel cell micro-CHP. In: Update Ene-Farm & Ene.Field Market Introduction Programs Webinar, 18 February 2016.

Tyler, R., 2014. CHP fuel cell durability demonstration (highly efficient, 5-kW CHP fuel cells demonstrating durability and economic value in residential and light commercial applications). Final report. Available from: <http://www.osti.gov/scitech/servlets/purl/1160148> (accessed 01.03.16).

US DOE, 2012. Hydrogen, fuel cells & infrastructure technologies program: multi-year research, development and demonstration plan (Section 3.4: fuel cells). Available from: <http://www1.eere.energy.gov/hydrogenandfuelcells/mypp/pdfs/fuel_cells.pdf> (accessed 06.06.16).

US DOE, 2016. Hydrogen, fuel cells & infrastructure technologies program: multi-year research, development and demonstration plan (Section 3.4: fuel cells). Available from: <http://energy.gov/sites/prod/files/2016/06/f32/fcto_myrdd_fuel_cells.pdf> (accessed 06.06.16).

Printed in the United States
by Publisher

Printed in the United States
By Bookmasters